wirth 12,90
093880

German Geographical Research Overseas

Supplementary Volume
to the Series

Applied Geography and Development

Edited by the
Institute for Scientific Co-operation,
Tübingen

German Geographical Research Overseas

A Report to the International Geographical Union

Edited by Eugen Wirth

Deutsche Forschungsgemeinschaft, Bonn
Institute for Scientific Co-operation, Tübingen

Edited by
Eugen Wirth

in cooperation with
Ernst Löffler and Manfred Schneider

Sponsored by: / gefördert durch:
**Deutsche Forschungsgemeinschaft
DFG**

ISBN 3-527-27014-0

Distribution:
VCH Verlagsgesellschaft mbH
Pappelallee 3, D-6940 Weinheim

Printed by Georg Hauser, Metzingen

Contents

Abbreviations 6

Overseas exploratory fieldwork – a specific tradition
in German geography
by Eugen Wirth 7

German geographical field research in the Western Pacific Region
by Hanns Jürgen Buchholz 27

Field research of German geographers in Latin America
by Jürgen Bähr and Erdmann Gormsen 51

German geographical research in Africa south of the Sahara
by Bernd Wiese 77

German geographical research in the Middle East and North Africa
by Eugen Wirth 93

German geographical research in the High Mountains of the World
by Carl Rathjens 133

German geographical research in the Polar Regions
by Hans Becker 149

Index 165

Dedicated to the memory of three outstanding German geographers whose overseas exploratory field work has set a standard and has been an inspiration to us all

During the first half of the 19th century:

Alexander von Humboldt (1769 – 1859)

During the second half of the 19th century:

Ferdinand Freiherr von Richthofen (1833 – 1903)

During the first half of the 20th century:

Carl Troll (1899 – 1975)

General abbreviations

Abh.	Abhandlung, Abhandlungen
Abh. Akad. Wiss.	Abhandlungen der Akademie der Wissenschaften
Arb.	Arbeit, Arbeiten
Abt.	Abteilung
Beitr.	Beiträge
Ber.	Bericht, Berichte
Diss.	Dissertation
Dt., dt.	Deutsch, Deutsche, deutschen
Erdkundl.	Erdkundliches
Erg.-Heft	Ergänzungsheft
Forschg.	Forschung, Forschungen
Geogr.	Geographie, geographisches
Inst.	Institut
Math.-Phys. Kl.	Mathematisch-Physikalische Klasse
Mitt.	Mitteilungen
Ser.	Serie
Univ.	Universität
Vol.	Band, Volume
Wiss.	Wissenschaftliche
Z. f.	Zeitschrift für

Abbreviations of periodicals

Afrika Kartenwerk, K. and B.	Afrika Kartenwerk, Karte und Beiheft
Coll. Geogr.	Colloquium Geographicum
Erdk.	Erdkunde
Erdwiss. Forschg.	Erdwissenschaftliche Forschungen
G. Helv.	Geographica Helvetica
G. R.	Geographische Rundschau
G. Z.	Geographische Zeitschrift
Geol. R.	Geologische Rundschau
Jahrb. G. G.	Jahrbuch der Geographischen Gesellschaft
M. G. G.	Mitteilungen der Geographischen Gesellschaft
P. M.	Petermanns Geographische Mitteilungen
TAVO	Tübinger Atlas des Vorderen Orients
Verh. … D. G.	Verhandlungen des … Deutschen Geographentages
Z. f. Geomorph.	Zeitschrift für Geomorphologie
Z. f. Geomorph., N. F. Suppl. vol.	Zeitschrift für Geomorphologie, Neue Folge, Ergänzungsband
Z. f. Erdk.	Zeitschrift für Erdkunde
Z. G. E.	Zeitschrift der Gesellschaft für Erdkunde

Overseas exploratory fieldwork – a specific tradition in German geography

by

Eugen Wirth (Erlangen)

Some years ago the President of the Association of German Geographers was to give a lecture in front of a fairly large audience of renowned scientists from other academic disciplines. The theme of the lecture was: "What is the significance of geography in Germany?" In the course of his preparation for the lecture he had already gathered many interesting facts and had followed several lines of development, but still found himself short of a central theme. Just for fun he asked me whether I could identify a longstanding characteristic feature which distinguished German geography from that of other Western countries. I was puzzled as to the answer – but then, after a little thought, I suggested such a feature: *Exploratory fieldwork overseas.* At first he hesitated but after a brief moment of deliberation he agreed. In order to clarify and justify this choice, it is necessary first of all to provide some background information to the topic.

Over the past three decades it was primarily academics from English speaking areas who were involved in the *development of new concepts within geography.* This new and later on worldwide development started in the *United States:* From 1951 some brilliant young geographers had been gathering at the University of Washington in Seattle to develop the concept of what was called "New Geography". During the time around 1970, the majority of the better known geography departments in the United States were dominated by representatives of the quantitatively orientated "Spatial Analysis". During the sixties this "New Geography" movement spread to *England. Swedish geographers* were also involved at an early stage in the development of new concepts; some of them worked at times in American universities.

At joint conferences in the USA, England and Lund (Sweden), these pioneers of "Spatial Analysis" repeatedly came together in order, through the close exchange of thoughts and ideas, to develop further and to consolidate the new approaches and methods. The studies, lecture halls and computer centres of the Western world provided the setting for this scientific work. Some occasional trips overseas were at best undertaken for the purpose of visiting the Statistics Offices there to collect figures for the computer. Fieldwork was of little significance for the modern model-building Geography.

The Anglo-Saxon countries and Sweden had also been the innovators in the formulation of further concepts which developed out of "New Geography"; in this context it is especially worth mentioning behavioural geography, perception studies, humanistic geography, phenomenology and radical geography. In the same way as "Spatial Analysis" these most recent geographical paradigms gained a foothold in the late seventies and early eighties at many universities in other Western states. The questions they dealt with, however, remained restricted to the post industrial society of the Western world. No interest was shown in the totally different worlds and societies overseas.

It is significant that a similar concentration of research interest in *German Geography* never occured. In the Federal Republic the development has taken another direction, conducted by the traditions of a specific German way. After the Second World War a long-standing sphere of interest was given new attention: *Extensive field studies* on specific topics, in both physical and human geography, in *other continents* and in *regions of other civilizations* were highly successful. This was the direction which the three great geographers Alexander von Humboldt, Ferdinand Freiherr von Richthofen and Carl Troll had pioneered and stimulated us. This line of geographical research was so important and so well regarded in Germany that it was able to maintain its status even in the face of the new approach that recommended the establishment of universal laws and the formulation of theories in the field of social sciences.

First place in a long line of German geographers whose life work was marked by involvement in overseas exploratory fieldwork is occupied by *Alexander von Humboldt* (1769–1859), the founder of Geographical Science. His five year long expedition to Latin America (1799–1804) led to many pioneering discoveries and new ideas. The results were recorded in a monumental work consisting of 35 volumes, entitled "Voyages aux régions équinoxales du Nouveau Continent, fait en 1799, 1801, 1802, 1803, et 1804 . . ." (Paris 1805–34). Humboldt had devoted six years to preparing himself for this journey, and needed more than thirty years for the analysis and evaluation of the results.

A second famous German geographer, *Ferdinand Freiherr von Richthofen* (1833–1905), spent almost twelve years from 1860 to 1872 travelling overseas. First he visited Ceylon, Java, The Philippines, Siam and Indo China. Then he did research work in California. Finally, between 1868 and 1872, he was able to make seven extended expeditions to regions of China's inland, which up to that point in time were still largely unknown. Subsequently, he worked for almost ten years on compiling the results of his explorations; these are published in "China. Ergebnisse eigener Reisen" ("China. The Results of my Personal Travels", 5 text and 2 atlas volumes. Berlin 1877–1912).

The life and work of *Carl Troll* (1899–1975) – surely the most distinguished German geographer of the last fifty years, and President of the International Geographical Union from 1960–1964 – once again reflects very clearly the relevance of overseas exploratory fieldwork in German geography. From May 1926 to autumn 1929 Carl Troll travelled around Latin America, primarily in the tropical Andes of Bolivia and Northern Chile, in Columbia and Ecuador. During the travels he drew up 10,000 kilometres of cartographic route records, scaled high peaks in the Cordillera Real, and collected a herbarium of 3,400 plants. H. BECK (1982, p. 276) speaks of an "Alexandrian campaign in modern Geography". The information gained about the countries as a result of the journey was published in the "Handbook of Geographical Science": "Die Tropischen Andenländer" ("The Tropical Andes Countries", Potsdam 1930). In 1933–34 there followed a second great expedition into tropical East Africa, and then in 1937 Carl Troll travelled to tropical Asia as the scientific leader of the German Nanga-Parbat expedition.

The field research and the observations carried out during these expeditions overseas formed the basis of the pioneering work which provided a comparative overall view of the climate and vegetation of the Tropics in its three dimensional zoning and landscape division (C. TROLL 1966, 1975). Alexander von Humboldt and Ferdinand Freiherr von Richthofen still represented unique exceptions within their own generation. Carl Troll, on the other hand, was regarded in German geography as primus inter pares; together with more than just a few other geographers of his generation he passed on the tradition of overseas exploratory fieldwork to the post war generation, who are still actively involved in their science today.

Naturally there was, in the *English speaking countries,* still a small group of geographers whose work dealt with foreign countries, peoples and cultures even at the time when the "New Geography" movement was thriving. However, they enjoyed less scientific prestige than the model-builders. Their treatment of faraway countries and states was often more "foreign area studies" than "overseas exploratory fieldwork". Some of them never even left their desks at home. Others who were resident abroad only worked in the libraries and archives there, or only obtained their information from the authorities and key people in the countries' capitals.

The difficulties which the *United States* was having in coming to terms with the concept of overseas exploratory fieldwork are shown in the 1973 anthology, published by the Geography Department of the University of Chicago: "Geographers Abroad. Essays on the Problems and Prospects of Research in Foreign Areas" (M. W. MIKESELL ed., 1973). Some of the authors in this anthology limit themselves to sketching the area on which their report is based in terms of a few characteristic aspects of information about the country. P. W. English, for example, does not go into the theme "Geographers Abroad" at all; he only describes the ecological trilogy of the Middle East (city, village and nomadism). Other authors make reference to "the uneasiness that many scholars now feel about foreign area research" and discuss "whether one can hope to achieve a thorough understanding of another culture". They establish "that discussion of the intellectual or scientific objectives of foreign area research had produced some disagreement" (p. 293 onwards).

The most convincing reports are those which include examination of the work done by geographers in other Western languages about the land, people, cultures, institutions, ecology and history – for example, the report by J. J. Parsons (Latin America) or C. A. Fisher (Southeast Asia). In his article on Latin America, which is situated in the close vicinity of the USA, J. J. Parsons is forced to conclude: "Our publication record is less impressive. There have probably been as many monographs on Latin American topics published by French and German geographers in the last decade as the combined English language output" (p. 45). Even in his introduction M. W. Mikesell had pointed out the fact that "Among German geographers especially, the tradition that university careers should be built on a foundation of prolonged field study in foreign areas is strikingly evident" (p. IV).

The *United Kingdom* is having similar difficulties. If we glance through those editions of the "Geographical Journal" which have appeared since the Second World War we do indeed find lots of reports about overseas regions. The authors, however, are not so much scientists as officers of the Colonial or Indian Civil Services, the Diplomatic or Consular Services recounting their observations and experiences. In his latest book "On Geography and its History" (Oxford 1986), D. R. STODDART includes one particular chapter, "Geography, Exploration and Discovery". In it he complains that for the Geographical Institutes in British Universities "the fieldwork movement came to be increasingly confined to local areas at home" (p. 146). Fieldwork consisted of excursions during which the academic teacher would show his students objects in the open country; any kind of observation, recording and interpretation in the field was limited to surveys of drumlins or questionnaires in supermarkets. "It is clear that such projects promoted the education and enjoyment of the participants rather than inspiring original contributions of knowledge" (p. 145).

Using the example of Joseph Conrad, D. R. Stoddart shows "the contrast between this world of action and discovery, and that of the lecture theatre and library" (p. 143). Then he poses two questions above all others: "Why is it so fashionable to denigrate and ignore our traditional concern with the farthest corner of the earth" (p. 143), and "why the practitioners of field observation came to be held in such low regard by the New Geographers of the universities and colleges" (p. 153). Following on from this he pleads a convincing case for exploratory fieldwork.

In this Stoddart makes a distinction between three grades of exploration and discovery: "First, and most simply, there is the inadvertent journey, often undertaken after some catastrophe under conditions of great hardship" (p. 148). "My second category is that of the planned journey, of which the aim is simply to proceed between known points, with no suggestion of adding to knowledge other than through traversing unfamiliar routes" (p. 149). "The systematic and organized accumulation of knowledge during planned investigations became the central objective and this forms my third category" (p. 150).

With this statement we have returned to the central theme of scientific geography in Germany which I referred to at the beginning of my article. It is *systematic and organized accumulation of geographical knowledge during planned oversea's fieldwork* which, already in the nineteenth century, characterized the scientific work of German geographers, for example Alexander von Humboldt or Ferdinand Freiherr von Richthofen. Therefore, when many German geographers during the later 1950s, the 1960s and 1970s went in for extensive field studies in other continents and regions of other civilisations, they were merely taking up once again a main focus of German geographical research which had been there a long time before. This tradition is, even nowadays, so firmly established in the minds of German geographers, that in many of the universities' Geographical Institutes field research outside Europe is regarded as a favourable precondition for a career as a university teacher.

It would be a fascinating task to examine in a lengthy historical, sociological and theoretical analysis of our science just how and why overseas exploratory fieldwork became a fundamental paradigm of German geographical research. Within the framework of our short introduction, only a few comments can be given in respect of this question.

In past centuries, Geography as a subject concerned with providing knowledge about the Earth had seen the *unveiling of that same Earth* – the exploration and discovery of unknown continents and countries – as its primary goal. In his inaugural lecture at Leipzig University, Ferdinand Freiherr von RICHTHOFEN (1883, p. 33) divides the history of discovery into three periods: A *first age of discovery*, in which exploration was concerned with the shipping routes of the great oceans, the spatial distribution of land and sea on our globe and the rough outlines of the continents. There follows from around 1750 to 1850 a *second age of discovery;* during this time the coastal pattern of the last of the still partly unknown continents, Australia and the Antarctic, were scouted and work began on recording cartographically the other large continents.

The *third* and the last *age of discovery* follows on from around 1850 onwards; this age tried above all else to eliminate the last blank spaces on the maps in inner Africa and Australia, in the mountains and central highlands of Asia, in tropical South America and in the Polar regions. Ferdinand Freiherr von Richthofen presents a very vivid portrayal of this last age of discovery (1883, p. 34). He tells of how the nations competed to send out explorers whose task it was to record new information and make it known to others, of a broad interest among the educated public in geographical knowledge about far-away lands and peoples, of the economic and political consequences of the expansion of our world knowledge, thus of European colonisation, world trade and the opening up of transport links with the distant areas. However, he also tells us that the reports and journey descriptions written at the time were in many cases unsystematic, naively descriptive and thus bore very little resemblance to geographical science.

During each of the three periods which Richthofen identifies within the history of exploration German researchers were only peripherally involved in the *unveiling of the Earth* in so far as this meant *heroic actions of exploration and discovery*. From the discovery of America and the first circumnavigation of Africa in the Fifteenth Century to the first expeditions to reach the North Pole, the South Pole or the summit of Mt. Everest in the Twentieth Century, it is mainly men from other countries whose names are recorded in the annals of exploration history: Initially they were men from the seafaring nations: Spain, Portugal, Holland, England. Later, Russians, Scandinavians and North Americans added their names to the list. Of course, the expeditions of Heinrich Barth or Georg Schweinfurth to Africa, of Philipp Franz von Siebold to Japan, or of Thaddäus Henke to South America can also be seen on the same international level. But the men who have achieved world fame, like Marco Polo, Vasco da Gama, Christopher Columbus, James Cook, David Livingstone, Fridtjof Nansen, Roald Amundsen or Sir Edmund Hillary are, quite rightly, the pride of other Western nations.

On the other hand, German scientists can be counted amongst the leaders at a very early stage if by *overseas exploratory fieldwork* we understand the quest for *systematic and organized geographical research* during planned investigations – that is, scientific research into uncovering causal and spatial interrelationships as well as investigating culturally and socio-economically distinct ways of life in regions outside Europe. The first reliable informations about Japan we owe to Engelbert Kaempfer (1651–1716), and it was Georg Forster (1754–1794) who, together with his father, wrote the famous account of Cook's voyage around the world 1772–1775. Carsten Niebuhr (1733–1815), a German working in Danish Service, has left us an astonishingly precise and informative picture of life in Southern Arabia. Niebuhr, like Kaempfer, prepared himself intensively over a period of years for his great journey. Carsten Niebuhr is probably the first ever European explorer in history to have fulfilled in every respect the strict requirements of thorough scientific preparation for an expedition overseas (H. BECK 1971, p. 92).

At a time when the interior of some continents was still totally unknown, *Alexander von Humboldt* (1769–1859) set new standards in the scientific exploration of worlds overseas. He was not concerned with eliminating blank patches on the map; the basic topography of the regions of tropical America which he travelled to between 1799 and 1804 were already known. This in itself, however, made it possible for Humboldt to carry out systematic and intensive preparations. "As the pioneer of the leading geographical conceptions for many decades, he brought together scientific geography and exploration in a unity up to that time undreamt of. A preparation period spanning six years was followed by the actual journey lasting five years. After this there came more than thirty years of evaluation and analysis, producing in the process the largest private travel account that history has to offer. The interplay of these three periods was called a triadic chord in the history of exploration. On its perfect harmony Humboldt's world fame was based" (H. BECK 1985, p. 227).

Humboldt's conception of overseas exploratory fieldwork, which is still valid today, was the one on which *Geography* was founded as a methodologically *modern empirical science*. With it he was well ahead of his time. At the beginning of the Nineteenth Century, so little was known about large areas of the Earth that the discovery of still unknown countries and thus the elimination of blank patches on the map seemed much more worthwhile. So Humboldt had no immediate successors; as had been the case in the preceeding centuries, so in the first half of the Nineteenth Century discovery remained the main aim of expeditions to distant lands. At best exploration meant the encyclopaedic assembling of material, accounts of noteworthy phenomena or references to utilizable raw materials and products.

More than half a century had to elapse before *Ferdinand Freiherr von Richthofen* (1833–1905) with his great China journeys (1868–1872) took up Humboldt's model once again and systematically developed it further. The strong influence of the viewpoints and interests of his age is certainly much more evident in Richthofen's case than it was with Humboldt, who was absolutely sovereign in his scientific conceptions. Accounts of the discovery of still wholly unexplored areas of the Earth were at that time meeting with a powerful response from a wide section of the public. Richthofen, too, was unable to escape this fascination for exploration and discovery.

The depth of Richthofen's attachment to both worlds – the old traditional one of heroic discovery and the new one of empirical fieldwork overseas – emerges from his speech at the 1878 Golden Jubilee celebrations of the Berlin Geographical Society (1928, p. 28f.): "So we see how exploration is making great strides to take hold of the Earth. Wherever an unknown stretch of land still exists, there are energetic men to be found to fill in the details about it. We are now witnessing a thrust forward in the battle for the last palms of victory still to be had in the opening up of the few wholly unknown regions of the Earth. The successful attempts to penetrate into unknown lands are actions which provide us with the finest examples of courage and energy, of bold venturing spirit and self sacrifice, of the spiritual superiority of the man of education and character. The weak are quick to give up. Only the strong attain their goals . . .

Indeed the areas of the Earth themselves are limited. We know the seas' coasts, and when, in the future, the five continents have been explored, the only unknown land left will be in the Polar regions. Like ice in the heat of the sun's rays, so the unknown regions melt away faced with the light of scientific knowledge. When the future generations look upon our lot, it will not be without envy, for it was granted to us to open up the world. Fortunately, the aims of science are not limited in so narrow a fashion. They expand along with the height of the viewpoint which has been achieved . . .".

Even among Richthofen's immediate students the conflict of strategies and aims still lingered on; *Alfred Tafel* (1877–1935) and *Sven Hedin* (1865–1952) both risked their lives on exploratory expeditions in order to make their contribution to the discovery of the last undiscovered areas of Tibet and Central Asia. *Alfred Philippson* (1864–1953), *Max Blanckenhorn* (1861–1947) and *Alfred Hettner* (1859–1941), on the other hand, travelled to areas of Asia Minor, Syria and the Andes, about which sufficient topographical information already existed, in order to carry out geographic field research in the true spirit of Humboldt's work. *Erich von Drygalski* (1865–1949) should also be counted in this second group of Richthofen students; during his great expeditions to Greenland and the Antarctic he intentionally renounced spectacular advances into still virgin territories in favour of an intensive scientific exploration of more easily accessible regions.

In his later years Richthofen himself clearly saw that it was not through discovery and exploration, but rather only through clear questions, concepts and methods that *Geography* would develop into a *modern empirical science*. From 1886 to his death in 1905 he worked in his position as Head of the Department of Geography at the University of Berlin to put this idea into effect. At the same time he was the President of the Berlin Geographical Society and the founder of the Institute of Oceanography. He was thus able to exert great influence and had highly talented students, many of whom were later appointed to professorial positions at German universities. To Richthofen research overseas was a must for every geographer; he inspired many of the oversea's expeditions of younger geographers and also supported them to the best of his ability. He even wrote a "Führer für Forschungsreisende" ("Guidebook for scientific exploration", Hamburg 1886).

Alexander von Humboldt was too far ahead of his own time. Thus Richthofen was the first scholar to establish overseas fieldwork as a specific tradition in German geography and to formulate clearly its approaches and its methodology: the observation of many individual components to form a causally and genetically based general picture; the conception of causal interactions between geo-factors as a dynamic process; the explanation of forms and phenomena as the result of dynamic forces and processes; the avoidance of random additive compilation of information in favour of a problem-orientated methodical research based on the observation of available facts.

Richthofen's successor in his Berlin professorship, *Albrecht Penck* (1858–1945), consistently carried the former's work further. After the First World War the age of great discoveries was over for good. It now became the indisputable task of geographical field research to concern itself with causal and genetic explanation and with the investigation of processes and sequences of development. A. PENCK himself writes on the subject (1928, p. 54): "The age of great geographical discoveries is over. . . . We are no longer dealing merely with the establishment of facts set down in a map, but, in accordance with the advanced level at which science finds itself, with the solving of problems. . . . Much more today than before observation goes hand in hand with the formation of sound hypotheses. The researcher must go to the correct place of observation . . . in order to be able to see and measure as many characteristic features as possible. His hypothesis guides him in this task; he rejects it if it is untenable, and replaces it repeatedly with new ones until the hypothesis agrees totally with the sum of the observations. In taking this path even those geographers working in narrow fields will be granted the joys of the explorer".

The main focus of Penck's empirical field research did not lie overseas, but rather in the Alps; we owe to him and Eduard Brückner the division of the Pleistocene into four ice ages separated by interglacial periods. However,

Penck learned about almost all the continents on long journeys and like Richthofen he provided lasting support for his students' fieldwork overseas; among Penck's most famous students we find *Norbert Krebs, Fritz Machatschek, Walter Behrmann, Hermann Lautensach, Hugo Hassinger, Herbert Louis, Julius Büdel.* Alongside them many other German geographers are now getting actively involved in overseas exploratory fieldwork.

Even before World War One Fritz Jaeger, Siegfried Passarge, Franz Thorbecke and Fritz Klute worked in Africa, Walter Behrmann in New Guinea, Karl Sapper in Middle America and Fritz Machatschek in Tienschan. After the war the trend grew even stronger; between 1919 and 1939 more and more German geographers travelled for months, occasionally for years, to other continents and regions of civilisation in order to carry out field research.

German geographers are now working overseas on *specific questions of Physical and Cultural Geography* which take into consideration the totally different and in each case peculiar character of foreign continents, climatic zones and areas of civilisation – questions therefore which at home in Central Europe could either not be worked on at all or not in an appropriate fashion. To name a few examples: the inland ice of the Polar regions, the origin of "Inselberge" (bornhards), the specific type of land forms in the deserts, the origin of the tropical tower karst, the characteristic features of tropical plantations, the way of life of the trecking Boers, European rural settlement and colonisation in the new world, the development of the cultural landscape of the South American Pampas.

Frequently, however, the long extended journeys undertaken in the years between the wars also produced material for *comprehensive regional syntheses.* Above all the series "Bibliothek länderkundlicher Handbücher" (since 1942: "Geographische Handbücher"), founded by Albrecht PENCK, provides synoptic accounts of land and people. Worth mentioning here are: Fritz MACHATSCHEK "Landeskunde von Russisch Turkestan" (1921), Wilhelm CREDNER "Siam" (1935), Norbert KREBS "Vorderindien und Ceylon" (1939), Albert KOLB "Die Philippinen" (1943), Hermann LAUTENSACH "Korea" (1945), Fritz BARTZ "Alaska" (1950). Some of these works bear the subtitle "written on the basis of personal travels and research". Works which in the same period appeared, although not in this series, are amongst others: Franz KÜHN "Argentinien" (1927), Leo WAIBEL "Die Rohstoffgebiete des tropischen Afrika" ("The raw material regions of tropical Africa", 1937) and Fritz JAEGER "Afrika" (1928).

The jubilee edition of the "*Zeitschrift der Gesellschaft für Erdkunde zu Berlin*" ("Journal of the Berlin Geographical Society") which appeared in 1928 to commemorate the society's centenary, shows clearly the preferential status of geographical fieldwork overseas. After an introductory historical article and two programmatic ones (by F. v. Richthofen and A. Penck), priority is given to seven contributions which report on fieldwork overseas. Only in a second part do we find articles on questions of European geography, on the geography of man made landscapes, on cartography, on oceanography etc.

The *conference reports and transactions of the German Geographers' Meetings* of 1921–1934 reveal the same trend. In 1927, for example, seven lectures about research overseas are given prime place at the beginning of the proceedings, and again in the 1929 volume six such papers are placed in priority position. In the two decades between the World Wars a lecture given at the Geographers' Meeting was viewed as a significant step towards a scientific qualification. In those days many famous German geographers gave conference lectures about their research in other continents and areas of civilisation; for example, Fritz Jaeger, Leo Waibel, Erich Obst spoke about Africa, Karl Sapper, Fritz Klute, Otto Maull, Hans Mortensen, Oskar Schmieder, Carl Troll, Franz Termer, Siegfried Passarge, Franz Kühn about Latin America, Ludwig Mecking, Heinrich Schmitthenner, Richard Finsterwalder, Arved Schulz, Hermann Lautensach about Asia, and Alfred Wegener and Herbert Knothe about the Polar regions.

During the time between the two wars there were also, of course, highly esteemed geographers working at German universities whose main study area lay in Central Europe. Along with Albrecht Penck we can also mention Hugo Hassinger, Robert Gradmann, Otto Schlüter, Joseph Partsch, Hans Schrepfer, Erwin Scheu, Hermann Wagner, Hans Dörries and Friedrich Metz. Most of them, however, were not envious of their colleagues working overseas; for these colleagues had chosen not only the more laborious, but also the more difficult area of work. After the First World War it was difficult for Germans to work in overseas territories in so far as they were under the control of other European powers. Besides, long journeys overseas demanded considerable personal finance, since the funds of the "Emergency Foundation for German Science" were severely limited. Hermann Lautensach, for example, sacrificed the total amount of his savings, some 8,000 Marks, for his journey to Korea. Those geographers who stayed at home then had a more comfortable existence at least with regard to their living conditions.

After the Second World War and the first difficult postwar years the tradition of overseas exploratory fieldwork was taken up again as quickly as possible in Western Germany. During the sixties and seventies field work overseas became even more widespread than before and the investigations reached a level of detail rarely achieved before. The preconditions for this development were very favourable: At the end of 1948 a total of 18 chairs of Geography in the Universities of the Federal Republic had been taken up once again; no fewer than fifteen of these eighteen professors had spent a long time working overseas! Furthermore, the students of those first postwar years belonged to a generation which had learned during the war to bear strains and to manage with little. During the Nazi regime this "war generation", to which this author also belongs, had been largely cut off from the rest of the world; we were now hungry for information about foreign countries and peoples, and felt the urge to get out soon and see everything with the own eyes.

When, in connection with the currency reform of 1948, the strict travel and foreign exchange restrictions were finally relaxed, geographers – professors as well as Ph. D. students – were among the first Germans to take up their scientific work in countries outside Europe again. Following the example of our academic teachers, obtaining our qualifications with fieldwork overseas became the guiding principle, the thing to be strived for, for most lecturers and assistant professors of my war and postwar generation. To mention just one example: About a dozen geographers wrote their doctorates under the supervision of *Carl Troll* in Bonn and then went on to work overseas and be appointed to professorial positions – Karl Heinz Paffen, Dietrich Hafemann, Felix Monheim, Wolfgang Weischet, Helmut Hahn, Wilhelm Lauer, Peter Schöller, Burkhard Frenzel, Ulrich Schweinfurth, Otto Fränzle, Dietrich Barsch, Hartmut Leser, Manfred Domrös.

However, quite a few of Carl Troll's students were also able to get to good positions in universities, regional institutions and planning authorities without having undertaken long journeys overseas. After the Second World War dynamic geographical research disciplines developed within the framework of German geography: an ecologically based differentiation and classification relating to the pattern of natural regions, a planning related Social Geography and Research into Central Places, a process oriented Agricultural and Settlement Geography, and a dynamic Regional Geography of post-war Germany. The followers of these branches enjoyed the same prestige as field researchers overseas and were taken into consideration in the same way when it came to job appointments. Those who, for the sake of personal scientific interests, found their fields of work in Germany or in Europe then faced no kind of discrimination.

In spite of this during the fifties and sixties the majority of the geographers of my generation decided to go in for exploratory fieldwork overseas. Even at this time the decision to do so still involved considerable sacrifices. The countries of the Third World, where the journeys led to, usually had no modern infrastructure, few hotels or other forms of accommodation, and drinking water was not always safe and medical care was poor. Also in most cases a considerable proportion of the travelling costs had to come out of the traveller's own pocket, and it was difficult to obtain a research permit in foreign countries. However, empirical field research overseas was a necessity for us if we were to answer many fundamental scientific questions, and it gave us an insight into new, totally different worlds. For these reasons we gladly accepted all the difficulties.

From around 1953 the *conference reports and transactions of the German Geographers' Meetings* reflect the status of exploratory fieldwork overseas. At this time it also becomes noticeable how the older generation of professors is being replaced by the students of the "war generation". The German Geographers' Meeting in *Essen* in *1953* organized a main session "Regional Studies and Cultural Geography Research Overseas". During it, geographers of the older generation gave reports about the USA, Canada, Columbia, Chile and India. In a second session "Morphological Research" papers were likewise given about oversea's fieldwork; however, alongside lectures from Professor Hans Mortensen about the USA and Professor Herbert Lehmann about the West Indies, there were also lectures from the two younger geographers Horst Mensching and Konrad Wiche on their research in Morocco.

The picture which emerges from the *Hamburg* Geographers' Meeting in *1955* is a similar one. In the representative main session "The present-day Geographical View of the World" lectures by established professors once again predominate. Herbert Louis delivered a lecture about Turkey, Albert Kolb about Southeast Asia, Julius Büdel about Ethiopia, Kurt Kayser about Africa, Hans Kinzl about Peru, Willi Czayka about Argentina. However, Carl Rathjens who gave a report on Afghanistan, must already be counted as belonging to the "war generation". In the sessions "Research and Observations during Travels" and "Cultural Geography" this younger generation – with the exception of Carl Troll's lecture about Central Mexico – almost stands on its own; Karlheinz Paffen gives a paper about East Brazil, Jürgen Hövermann about Ethiopia, Wolfgang Pillewizer about the Himalayas, Eugen Wirth about Iraq and Hans-Günter Gierloff-Emden about Texas.

The lectures given at the *Würzburg* Geographers Meeting in *1957* clearly reveal the increasing specialization of German geographical fieldwork overseas: One main session "Planation Processes and Pediplanation in the Humid Tropics" deals with problems of Physical Geography of an oversea's climatic zone; another main session "The Industrialization of Raw Materials Exporting Countries Overseas" looks at problems of economic geography of the Third World. Geographers from the older and the younger generations are just about equally represented here.

Finally, let us turn away from the observation of the German Geographers' Meetings of the immediate post war years back again to our more general perspectives. If we ask the more than 90 geographers who were working as heads of departments at universities in the Federal Republic during the two decades between 1949 and 1969 to name the *main focus of their research*, 70 per cent are distinguished by fieldwork overseas. The remainder are divided into two groups, each of 15 per cent: the one having worked predominantly in Europe outside Germany, the other having worked either only in Germany or alternatively on general geographical topics without any specific regional focus.

The figures quoted by E. LICHTENBERGER (1980, p. 21) point in the same direction. Between 1960 and 1975 two thirds of the individual geographical projects sponsored by the German Research Foundation are said to have been related to research plans outside Europe. Frau Lichtenberger is wrong though when she says that the preferential treatment given to such work was an intentional part of the German Research Foundation's research policy. The Foundation does not ask about the region which has been singled out for the project, but rather only about the quality of researcher and project. Therefore, the main areas of sponsorship approval are a true reflection of the main areas of the research projects chosen by the applicants themselves.

A more careful analysis of the figures shows that Frau Lichtenberger's results have to be revised. They are, however, still impressive enough. From 1960 to 1975 the *German Research Foundation* sponsored in its normal application procedures 1,412 individual geographical projects. Out of these 17 per cent were related to Germany, 23 per cent to Western Europe, 3 per cent to the Socialist Countries and 47 per cent to regions outside Europe. The last 10 per cent of the projects dealt with topics which could not be classified regionally (U. FAR-HOLLENDER and E. EHLERS 1977/78, p. 257). Furthermore, the *Volkswagen Foundation* also provided substantial and wide-ranging support for German geographical research overseas: 25 projects in the programme focusing on the "Near and Middle East" (1970–1981), 19 in the "Southeast Asia" programme (1976 onwards), 9 projects in the programme "Recent Developments in Latin America, Asia and Africa" (1981 onwards) as well as 4 projects in each of the two past projects "Research in Latin America" (1964–1975) and "Research in East Asia" (1964–1975).

Alongside the many research projects of German geographers financed by the German Research Foundation or the Volkswagen Foundation there exists a presumably large, but not precisely determinable, number of projects which have been financed by the scientists themselves out of personal funds, by their university out of its funds, from the funds of organizations for developmental aid, peace research etc. This considerable expenditure of money, but also of enthusiasm, personal involvement and idealism has yielded rich dividends. A large proportion of our extensive German field studies in other continents and regions of foreign civilization during the 1960s and 1970s were highly successful; the scientific community of German geographers has fully appreciated these research projects and their respective researchers.

Before this background we begin to understand why the paradigms of the English speaking "New Geography", Spatial Analysis and Regional Science, did not catch on in Germany. The beginnings for it were undoubtedly present. Led by *Dietrich Bartels,* a dynamic group of young German geographers has been demanding since 1968 that human geography should be conducted in line with a modern, theory oriented social science. On the model of English-language geography all geographical research ought to aim itself at universally valid laws or comprehensive theories of social science.

However, the challenges of Dietrich Bartels were only followed by a relatively small group of German geographers. On top of this, their contributions to the "New Geography" met with a very critical response: Quite often the Anglo-Saxon concepts and scientific fashions are adopted and imitated uncritically in German geography. Sometimes it was no longer the independent and profound thinker who enjoyed the best reputation among the co-modernists, but the person who managed to take over and imitate "modern" procedures most skillfully, and to present them most convincingly (cf. E. WIRTH 1981).

Also the paradigms themselves of English-language geography – geography as a theory oriented social science, phenomenology and humanistic geography as hermeneutic approaches – were subjected to fundamental criticisms. At the same time as all this was happening German geographers were putting their efforts into laying the epistemological and theoretical foundations of geographical research overseas (cf. E. WIRTH 1984). Can overseas exploratory fieldwork be justified on the basis of *scientific theory,* or do we continue to see it as deriving from the *history* of German geography? And how do we account for its *sociological aspects?*

In a stimulating paperback of modern social anthropologists called "Feldforschungen" (Field Research, H. FISCHER ed. 1985) reference is made to the fact that exploratory ethnological research in overseas countries can become a *ritual* and an *ideology* – field research finds its value within itself, it is only continued for its own sake: "There are too many students who dream of field research, of becoming explorers, too many ethnologists who feel happiest carrying out field research simply as field research for its own sake and who make no mention of the context of the research aims, the problems experienced, the results" (p. 9).

Justin STAGL (p. 289 ff.) compares the personal experience of field research overseas with a pilgrimage full of sacrifices, with an adventure story or a novel tracing the development of a character, with the fortune of a travelling knight in a strange land: "In Ethnology the field researcher is the same as the Mecca pilgrim in Islam. This pilgrimage . . . is the duty of all believers. Only those who have been on it are Muslims in the full sense of the word; they may bear the title 'Hadschi'. Field research can be seen as . . . a personality forming experience, the actual purpose of which is to make a young person with ethnological knowledge into a true ethnologist. It is a test of skill, in which he can succeed, or on which he can founder . . . The transformation of the personality through participation in field research is regarded as one of the main preconditions for membership of the ethnologists' guild. This is, therefore, something like an order made up of those who have been put to the test and have been initiated".

Alongside such a personal and individual aspect of field research as ideology, Stagl also points to another collective aspect (p. 302 ff.): If the ethnologists' guild can cast aside the "easy chair ethnologists" as "not having proven themselves scientifically", thus keeping out any unwanted applicants, it improves the work possibilities and job prospects of their members who have undergone the test of field research. In view of the limited resources available, therefore, the guild is kept down to a few members. "Of course, Ethnology, being blessed with heroic legends about its founders and with a strict and demanding initiation ritual, was more than a mere guild; it was a close alliance . . . But speaking in strict confidence, it has to be said that field research today is no longer what it was earlier. . . . The student who sees field research as the springboard into a career, transforms it from a secular pilgrimage and a spiritual adventure into a exercise of duty which has to be completed in the interests of a job. The professor who may even already stipulate it as a task for his students should not be surprised if its substance – and unfortunately its aura too – gets lost along the way".

With geographical field research overseas there is undoubtedly less danger of such an individual or collective ideologization than is the case with the ethnologists. Nevertheless, we geographers should consider very critically the *sociological aspects of our scientific identity*. Field research overseas is never commendable for its own sake; it only has value as a way to achieving a goal, which cannot be achieved in any other way – a goal for the benefit of our Western Society as well as of the Third World.

For the geographer empirical field research in other continents, climatic zones and cultural realms is necessary quite simply because *the world of our highly developed post industrial Western society does not constitute the whole world.* In the well renowned books of the English language "New Geography" we repeatedly find general statements about "human beings"; however, these statements are in the end only valid for members of the American Middle Class. Only empirical fieldwork overseas gives a picture of the *scale of variation in this Earth's structures and processes*, as well as of the *breadth of variation in the strategies of human activity and the contexts of human decisions.*

From this we can draw the conclusion that all general statements about the Earth, about people and about the relationship between man and his environment can only claim a validity which encompasses the whole globe when they have been tested and supported by means of empirical research overseas. On top of this, our picture of this world will only achieve some degree of accuracy when we come to understand, through empirical fieldwork overseas, the wholly different nature of the conditions in other continents, climatic zones and cultural realms. Indeed, it is repeatedly stressed that at the turn of the Twentieth to the Twenty-first Century the different parts of the Earth will have grown together to form "one world". On this point, however, we should not forget that the regional peculiarities and spatial distinctions of this "one world" will still be there in the future. It is Geography's task to investigate them.

In the course of the coming decades the central focus of the "one world" will shift considerably. In the year 2030 it could be the case that only 15 per cent of the world's population still lived in the Western and the Socialist industrialized countries, with the other 85 per cent living in the developing countries. In the future much more than in the past, it will be the geographer's task to direct his attention towards *world wide interdependencies* and in particular towards the *situation in the countries of the Third World.* This, however, is only possible with the aid of geographical field research overseas.

Scientific work in one's familiar study room in the restricted environment of one's homeland or in areas of related culture and language is undoubtedly more comfortable and less arduous than exploratory fieldwork overseas. Even if the large German research foundations make available more funds than corresponding organizations in many other states, overseas fieldwork generally requires noticeable financial sacrifices as well. However, geographical science itself would give up if geographers were not repeatedly prepared to gladly take on, such as they are, the difficulties of overseas exploratory fieldwork.

For many decades German geographers have developed and maintained strong ties of cordial friendship with the peoples and countries overseas. Geographical research would loose all meaning if it did not contribute in some ways, directly or indirectly, to the improvement of conditions in the countries researched. For instance geographical work provides a solid foundation on which politicians and planners in the Third World can base their decisions. The publication of the following survey is intended to be a small indication of gratitude and an expression of the close bonds German geographers have developed with their colleagues overseas. Without the active and generous help and co-operation of the local experts, colleagues, administrative departments and firms, it would be impossible for us to conduct our research there.

Bibliography

BECK, H.: Grosse Reisende. Entdecker und Erforscher unserer Welt.
München 1971.

BECK, H.: Grosse Geographen. Pioniere – Aussenseiter – Gelehrte.
Berlin 1982.

BECK, H.: Die Geographie Alexander von Humboldts.
In: Alexander von Humboldt. Leben und Werk.
Ingelheim 1985, p. 221–238.

FAR-HOLLENDER, U. AND E. EHLERS: Geographische Forschung in der Bundesrepublik Deutschland und die Deutsche Forschungsgemeinschaft, 1960–1975.
Geogr. Taschenbuch 1977/1978. Wiesbaden 1977, p. 241–253.

Festschrift Ferd. Freiherrn von Richthofen zum 60. Geburtstag am 5. Mai 1893, dargebracht von seinen Schülern.
Berlin 1893.

FISCHER, H. (ed.): Feldforschungen. Berichte zur Einführung in Probleme und Methoden.
Berlin 1985. (Ethnologische Paperbacks).

HUMBOLDT, A. VON: Voyage aux régions équinoxales du Nouveau Continent, fait en 1799, 1800, 1801, 1802, 1803 et 1804 . . .
35 vols., Paris 1805–1834.

LICHTENBERGER, E.: Zur Standortbestimmung der Universitätsgeographie. Reflexionen über die institutionelle Situation in der BRD und in Grossbritannien.
Mitt. Österreichische G.G. 122 (1980), p. 3–48.

MIKESELL, M. W. (ed.): Geographers abroad. Essays on the problems and prospects of research in foreign areas.
Chicago 1973. (The Univ. of Chicago, Departm. of Geography, Research Paper 152).

PENCK, A.: Neuere Geographie.
Z.G.E. Berlin 1928, p. 31–56. (Sonderband zur Hundertjahrfeier der Gesellschaft).

RICHTHOFEN, F. VON: China. Ergebnisse eigener Reisen.
5 vols. Text and 2 vols. Atlas, Berlin 1877–1912.

RICHTHOFEN, F. VON: Aufgaben und Methoden der heutigen Geographie.
Leipzig 1883.

RICHTHOFEN, F. VON: Führer für Forschungsreisende.
Berlin 1886.

RICHTHOFEN, F. VON: Triebkräfte und Richtungen der Erdkunde im neunzehnten Jahrhundert.
Z.G.E. Berlin 1903, p. 655–692.

RICHTHOFEN, F. VON: Die Geographie im ersten Halbjahrhundert der Gesellschaft für Erdkunde.
Z.G.E. Berlin 1928, p. 15–30. (Sonderband zur Hundertjahrfeier der Gesellschaft).

STODDART, D. R.: On Geography and its History.
Oxford 1986.

TIESSEN, E. (ed.): Meister und Schüler. Ferdinand Freiherr von Richthofen an Sven Hedin.
Berlin 1933.

TROLL, C.: Die Tropischen Andenländer (Bolivien, Peru, Ecuador, Columbien und Venezuela). Curaçao, Trinidad und Tobago. Guayana.
In: F. Klute (ed.): Süd-Amerika in Natur, Kultur und Wirtschaft.
Potsdam 1930, p. 309–481. (Handbuch der Geographischen Wissenschaft).

TROLL, C.: Ökologische Landschaftsforschung und vergleichende Hochgebirgsforschung. Wiesbaden 1966. (Erdkundl. Wissen 11).

TROLL, C.: Vergleichende Geographie der Hochgebirge der Erde in landschaftsökologischer Sicht. Eine Entwicklung von dreieinhalb Jahrzehnten Forschungs- und Organisationsarbeit.
G.R. 27 (1975), p. 185–198.

WIRTH, E.: Kritische Anmerkungen zu den wahrnehmungszentrierten Forschungsansätzen in der Geographie. Umweltpsychologisch fundierter "Behavioural Approach" oder Sozialgeographie auf der Basis moderner Handlungstheorien?
G.Z. 69 (1981), p. 161–198.

WIRTH, E.: The abolition of man in modern social science orientated geography? New directions for rediscovering man and culture in geography beyond the "behavioural" and "humanistic" approaches.
London 1984. (University of London King's College. Department of Geography. Occasional Paper 23).

German geographical field research in the Western Pacific Region

by

Hanns J. Buchholz (Hannover)

1. The time before World War Two

It is a tradition of German Geography to carry out research not only in Germany and in the immediately surrounding areas, but also in foreign cultural regions and preferably amongst various groups of society. The general aim is to enlarge the spectrum of scientific knowledge, to study strange phenomena and to import the knowledge on foreign structures and development processes. Only a few of the large number of German scholars who have conducted research as early as in the 19th century and especially since the beginning of the 20th century in the countries of the Western Pacific Region, may be mentioned here.

Ferdinand von RICHTHOFEN visited *China* in 1860 and more extensively between 1868 and 1872 and published his significant China texts in combination with an atlas. Following World War One and the foundation of the Republic of China, the interest of German geographers in China grew rapidly: on the occasion of various tours Heinrich SCHMITTHENNER mainly investigated regional and urban geography; Friedrich OTTE dealt with questions concerning economic geography. Hermann von WISSMANN, professor at the University of Nanjing 1934–35, was engaged in regional studies of physiogeography and of climatically controlled land forms. Wolfgang PANZER's research is based on the geomorphology of South China; he was professor at the Sun Ya-Tsen University at Guangzhou/Kanton from 1931–1934.

The publication of an extensive regional geography of *Japan* based on the research work of Johannes Justus REIN started in 1881. In the 1920s Ludwig MECKING studied cultural geographical processes in Japan, particularly the development and importance of Japanese towns. His book on "Japanese Harbours" (1931) and his essay on "Japanese Urban Landscapes" (1930) provide important insights into the basic structure of Japan even today. Martin

SCHWIND, whose research work as a result of World War Two was mainly published from the 1950s onwards, started his studies in Japan in 1934 (first on glacial morphology). Also to be mentioned is the Austrian geographer Leopold SCHEIDL (Vienna) and his geographical research work on settlements in prewar Japan.

Many geographers of that era tried to write a classic extensive *regional geography* – such as Hermann LAUTENSACH (1945) on Korea, Albert KOLB (1942) on the Philippines, Wilhelm CREDNER (1935) on Thailand, Walter GEISLER (1939) on Australia, New Zealand and the South Pacific – always based on a number of case studies. Furthermore, a great many German scholars dealt with detailed problems, or – in countries lacking a geographical science – endeavoured to "unveil" unknown regions: Karl HELBIG (1931) researched geographical aspects concerning Indonesian settlements, Herbert LEHMANN (1936) travelled to Java, Sumatra, Bali, Flores and Celebes and published several articles on the population and the colonial urban development. K. DOMIN (1928) conducted studies in vegetation geography in Australia, and Karl SAPPER (1915), Walter BEHRMANN (1928) and G. FRIEDERICI (1912) took stock of regional geographies in the South Pacific. Concerning this vast and polymorphous island region, particular credit must be given to the splendid ethnological research work of Augustin KRÄMER (1927) and other members of the great South-Sea Expedition (1908–10) of Georg THILENIUS (1927).

Much of this research proved to be rather adventurous and usually lasted some months or even years. The lengthy journey from Germany had to be undertaken by ship. Transportation problems inside many South East Asian and South Pacific countries were even greater, due to the fact that these areas were actually "unknown". Research trips had the character of *expeditions.* Additionally, no exact maps could be obtained. Thus one of the first priorities of such expeditions was to draw an appropriate topographic map. In East Asian countries official permits were necessary for such research tours, but on the whole, they were mainly granted generously in the colonies of South East Asia; or the South Pacific research work was hindered by communication problems with the native population, which often proved to be quite dangerous. Nevertheless, almost all states and territories in the Western Pacific region were repeatedly been visited by German geographers, especially after the end of the 19th century.

The Second World War caused a significant break in the development of German geographical research work in the Western Pacific region. For more than a decade war and its aftermath had put an end to all research projects. The lack of financial funds in the early days of the Federal Republic of Germany, the gradual cessation of hostilities between Germany and a number of nations, as well as the occupation of Japan, the policy of isolation by the communist PR China, the Korean War, the Vietnam War and the entire post-war situation resulted in cautious and timid research trips in the 1950s.

2. East Asia

Martin Schwind was the first to re-establish scientific links between German geographical research and *Japan*, which were then developed and formed by Peter Schöller. In 1956, Martin SCHWIND visited the Far Eastern island region for the first time after the Second World War. Impressed by the changes due to modernization, industrialization and urbanization, he began to link the Japan of the 1920s to the Japan which presented itself to him at the time by interpreting the structures and developments of the country as an objectification of the mental tradition and basic principles of Japanese society. In 1967 his first volume (the Natural Landscapes) and in 1981 the second volume (Cultural Landscapes. Economic Power on Limited Land) as part of his extensive regional geography were published.

Peter SCHÖLLER conducted a number of research tours in Japan from 1957 onward with the support of the "Deutsche Forschungsgemeinschaft" (German Research Foundation). His studies concentrate on urban structures, town development and urbanization processes and can therefore be regarded as an analysis of the most significant feature during the present Japanese era. The connection of detailed empirical surveys with nation-wide structure and development data as well as with spiritual and cultural historical facts show the advantages of real geographical analysis. Neither the interpretation of abstract statistics nor the evaluation of secondary literature could have resulted in such fundamental research findings as presented by Schöller. Due to his and Teiji Yazawa's (Tokyo) initiative, German-Japanese research work-seminars have been held at regular intervals in Germany and Japan since 1965, introducing younger geographers of both countries to research work in the respective country.

Publications on centralization and centre shifting, on growth and changes of Japanese urban regions, on migration and town development, on the enlargement of underground centres and so on have also influenced the scientific geography of Japan and have partly been translated into Japanese. Recent studies deal with new urban developments due to tourism, due to regional planning measures ("technopolis") and to the influences of new religions (P. SCHÖLLER 1960, 1962, 1969, 1973, 1976, 1980, 1984). In 1978 Schöller edited a regional geography of East Asia, in which he contributed the regional sections on Japan, Hong Kong and Macau.

His student, Winfried FLÜCHTER (1975, 1978, 1985), has been conducting research in modern urbanization and national development processes with special consideration of questions of applied geography: the acquisition of new land, industrial settlements, town planning and ecological problems.

To foreign scientists Japan presents specific difficulties as a result of their usually limited knowledge of the Japanese language. This does not so much apply to their own research work, because some English is spoken all over Japan and any difficulty may be overcome with the assistance of an interpreter. But it is of major importance that Japan has its own manifold and developed geographical science and therefore scientific studies are usually published in Japanese. Without the knowledge of the Japanese language, this literature cannot be used, which results in a considerable limitation of research activities by German geographers in Japan. W. Flüchter is one of the few German geographers who speaks Japanese.

The topics of German research work in Japan are numerous. Gerhard AYMANS (1965, 1980) has studied matters of the economic and cultural geography of the Japanese aqua-culture in detail. Later on, he researched demographic anomalies which have been affected by Japanese tradition.

In general, a number of foreign geographers found interest in typically Japanese phenomena, often connected with aspects of tradition. Karl-Emil FICK (1984), for example, interpreted Japanese garden arrangements as an expression of the Japanese structure of society. The Swiss geographer Hans BOESCH (1957, 1977) analysed traditional Japanese land utilization. He also commented on problems concerning urbanization and traffic in Japan, as did Hans HILLENBRAND (1980, 1983). An interesting study was undertaken by Alois MAYR (1984) on the system of Japanese inland air-traffic. In physical geography, the only German researcher was Ludwig ELLENBERG (1974, 1976, 1982) who studied the climatic effects on the development of land forms.

Until the end of the 1970s research work by foreign scientists was impossible throughout the PR China. It was also remarkable that neither Taiwan, Hong Kong nor Macau attracted any German geographer for a long time. Certainly one has to take into consideration that the *Republic of China (Taiwan)* is still under martial law. Nevertheless, apart from a short study on the structural changes of Taiwan by Otto TIMMERMANN (1963) only two other geographical research projects were carried out by Wilfried KOCH (1971) on the subject of structural changes on the town periphery of Taipei and by Karl-Heinz WERNEBURG (1980) on export processing zones. Of course, a variety of scientific texts on Taiwan have been published, but they were not based on actual fieldwork – with the exception of the above mentioned.

The same applies to the *PR China.* However from the 1960s onwards, a gradually increasing touristic activity involving German geographers can be recorded. But even in the 1970s China-travellers were classified as exotic. Their books and publications aroused great interest, although the authors were usually allowed to visit only a specific sample of locations. Therefore, information on a restricted number of towns, communes and dams repeatedly appeared in this literature. The regional geography of China by Albert KOLB (1963), too, was written without field research in that country.

The relative opening of the PR China since 1971/72 has encouraged many German geographers to visit this significant nation, which had been hidden so mysteriously until then. But "research work" has not been conducted. People have tried, rather, to gather information by means of conversation and observation in order to begin to comprehend the structures and developments of this country. In the mid 1970s Johannes KÜCHLER (1976) was still not able to pursue actual research work, but had to conduct his highly interesting studies on the structures of China's late town development whilst on a prolonged stay, which had not been scheduled as a research project.

Only since the beginning of the 1980s, has the number of German geographers in the PR China been increasing. In relation to the size and complexity of this country, the number of German scientists who actually work in the PR China is extremely small. There are some reasons for this. On the German side, there are hardly any geographers capable of speaking or reading Chinese. On the Chinese side, one will find a very limiting selection of permits for scientific research programs. To some extent this may be explained by the restricted hotel and accommodation capacity. More substantial, however, is another fact: the Chinese authorities are only vaguely interested in the numerous cultural geographical projects, unless present questions of planning are concerned. Finally, it is a handicap that even on the basis of a good knowledge of the Chinese language a profound analysis of geographical phenomena is hindered, due to the lack of scientific publications by Chinese geographers. This is partly a result of the secrecy of many analyses, but it is also particularly due to the fact that practically no scientific research took place during the terrible so-called "cultural revolution" and shortly afterwards.

Under these restrictive conditions Heiner DÜRR (1983, 1984) was engaged in research concerning matters of political and population geography and together with Urs WIDMER he investigated the field of disparities in rural areas. Norbert WEIN (1985) analysed the agriculture in the Wei-Hi-Basin in detail. Jörg STADELBAUER (1984) worked on population problems and Wolfgang TAUBMANN (1985, 1986) conducted research on inner-town development in 1981/82. The physiogeographical research work in China is limited to one very interesting study on altitudinal zonation of landforms in the Tibetian

Himalaya by Matthias KUHLE (1985). All in all, the German empirical research activities mainly concentrate on very few and very limited sites in the PR China.

This applies even more so, to an even greater extent, to the *Korean peninsula*. In the Democratic PR Korea (North Korea) no research by German geographers has taken place since World War Two. Regarding South Korea one can only mention the German geographer Eckhard DEGE (1979, 1985), who often and thoroughly examined agro-geographical structures and developments and has recently engaged in questions of population distribution in South Korea.

In *Hong Kong* Peter SCHÖLLER (1967) began his research in 1959. He experienced the British Crown Colony as a giant refugee camp and wrote his first comprehensive analysis from the viewpoint of social and political geography. Inspired by this, Hanns J. BUCHHOLZ conducted empirical research on problems of population development, the extreme population concentration and on problems concerning population mobility and settlement behaviour. A number of research visits, some together with Peter Schöller, resulted in the publication of a broad analysis on Hong Kong as a development pole for the change of China in 1985. Research activities in Hong Kong's surrounding areas had preceded, i.e. the Special Economic Ẓone Shenzhen and the Portuguese colony Macau (Hanns J. BUCHHOLZ 1973, 1978, 1985, 1986; Peter SCHÖLLER 1984).

Furthermore German geographers investigated a number of detailed problems: Johannes KÜCHLER (1971) studied spatial disparities, Otto J. GOLGER (1972) worked on accomodation and urban renewal, Klaus FRIEDRICH and Reinhard HELMSTÄDT (1985) on the German population in Hong Kong. Dusan SIMKO (1983) from Switzerland worked on the hawkers of Hong Kong.

3. South East Asia

The postwar situation in the countries of South East Asia was – similar to East Asia – rather intricate: in Vietnam civil war broke out in 1948, which eventually led to the Vietnam War and involved the neighbouring states of Laos and Cambodia. In Cambodia no consolidation has yet taken place. Accordingly, no one has been able to conduct any research. This also applies to Burma (independent since 1948), whose socialist government imposed a relatively strict policy of isolation.

Thailand, Malaysia, Singapore, Indonesia, the Philippines and Brunei (British until 1985) offered more promising working possibilities for foreign scientists. Harald UHLIG has an international reputation for his South East

Asian research (1975, 1979, 1980, etc.). Since the end of the 1950s, he has carried out a great many research projects in South East Asia and has presented a large number of publications on this region. Following initial studies on ethno-social structures on the Malayan peninsula, he extended his work onto the northern areas of Borneo and, furthermore, onto exemplified areas of Indonesia. His main interest was and is the connection of social and agro-geographical questions. Of great importance, among others, were his articles on the economic and social geographic differentiation of the altitudinal zones in the mountain states of South East Asia (and the Himalaya). A further research interest was based on rice cultivation. Since the end of the 1970s, Uhlig has been engaged in surveys of spontaneous and planned land colonization starting in Southern Thailand.

On many occasions his students have either assisted or continued his research work. Johannes KÜCHLER (1968) wrote his doctoral dissertation on ethno-social forces, which have changed the cultural landscape on Penang (Malaya). Ulrich SCHOLZ (1977, 1980) investigated land utilization systems in Sumatra (Indonesia) since the beginning of the 1970s and presented his doctoral dissertation on the subject of the agricultural structure of West Sumatra and its development possibilites. Studies concerning tropical shifting cultivation and altitudinal zonation from the viewpoint of physio-geography and especially of vegetation geography by Norbert STEIN (1972, 1974) date from the same period. He also worked in Sumatra and later on in the Western Malayan archipelago. Sumatra and Java (Indonesia) are also the research areas of Manfred DOMRÖS (1976), who studied the climatic influences on agriculture.

In general the *structure and development of agriculture* have been the main topics of most research projects by German geographers in South East Asia. The concentration on these topics is quite understandable, because the development process of South East Asian countries strongly depends on the unfolding of agricultural resources. As a matter of fact, a number of German geographers – apart from those already mentioned – have studied similar topics. Herbert WILHELMY travelled to Indonesia (1966/67), to the Philippines (1972/73) and to other countries in South East Asia and presented various trend-setting publications on rice cultivation and food problems (1975). Gotthard BURGER (1975, 1976) investigated agricultural intensifying measures in Java (Indonesia) and especially studied the interrelation between diseases and the manner of nutrition in these regions. Klaus HAUSHERR (1968/69) examined agro-cultural problems in the Philippines. Jürgen HOHNHOLZ (1986) conducted extensive research work on the subject of agricultural production in Thailand. Wilhelm MATZAT (1973) also studied rice growing and the village structure in Northern Thailand.

Wolfgang SENFTLEBEN (1971), at that time working as a scientist in Taiwan, investigated (apart from other aspects) the tenureship-policy of Malaya with regard to agriculture. Joachim METZNER (1977) studied the agricultural development with the consideration of geo-ecological aspects of various insular states of Indonesia – East-Timor, Sumba, Flores – and developed an excellent analysis of environmental problems in East Timor. And last but not least, Werner RÖLL (1979, 1981, 1983) has published on the agricultural land tenure and on soil utilization in Java (Indonesia) or, together with Gerd ZIMMERMANN (1979), in Western New Guinea (Irian Jaya). There are other agro-geographical studies by Zimmermann for Indonesia (1980). Werner Röll also has investigated population and settlement geographical aspects of Indonesia since the beginning of the 1970s, being especially interested in questions of population dynamics and population distribution. He also wrote a regional geography of Indonesia (1979).

Other research projects concerning *population geography* were conducted by Kurt HORSTMANN (1964), Gerd ZIMMERMANN (1975) and Werner RUTZ (1977) in Indonesia and, to a certain extent, by Dietrich KÜHNE (1969) in Malaya. In this context, the study by Albert KOLB (1974) on foreign minorities – Chinese, Japanese and Indians – on the Philippines should be mentioned, although the social geographical or, in general terms, the cultural geographical handling of this topic exceeds population geographical aspects.

In relation to *urban geographical research*, there are only few German geographers to be mentioned, e.g. Hans-Georg GLAESSER (1976) on Bandung (Indonesia), Elmar FICHTER (1972) on Jakarta (Indonesia), Johannes KÜCHLER (1971) on Singapore and the numerous research tours – dating back to the pre-war period – and comprehensive studies on Manila (Philippines) by Albert KOLB (1975, 1976), partly in co-operation with Jürgen EHLERS and Frithjof VOSS. One has to emphasize the thorough research by Dietrich KÜHNE (1976) on urbanization processes in Malaysia. Kühne visited Malaysia often and at length and especially analysed Greater Kuala-Lumpur. A very detailed description of Indonesian towns was completed by Werner RUTZ (1985) who had already written a book on the development of traffic routes among the outer Indonesian islands (1976). Friedemann CORVINUS (1984) traced the specific problem of the energy supply of Malaysia in his dissertation.

In Indonesia, the Philippines and lately also in Malaysia *regional planning* and country development have become a part of the German geographic research work. Heiner DÜRR (1977) worked on regional planning for West Sumatra, likewise Ulrich SCHOLZ (1980) for Central and South Sumatra. Dirk BRONGER (1983) studied regional disparities, structures and problems of the metropolization and chances of special planning arrangements for the development of the Philippines. In the mid 1980s Elmar KULKE (1986) and Knut

KOSCHATZKY (1986) investigated the connection between industrialization and polarization in the development process of Malaysia in two separate examples. Johannes C. FRANZ developed a well researched and very interesting regional study on the Sultanate of Brunei in 1978.

In general one can say that almost all German research work in South East Asia is based on cultural and applied geographical topics. Research work on *physical geography* is rare. Apart from the already mentioned studies on agrogeography by M. Domrös (climatology), N. Stein (geo-ecology) and J. Metzner (geo-ecology) there are only two research projects worth mentioning: a study on vegetation geography in the mountain regions of Java by Ulrich SCHWEINFURTH (1974) and the geomorphological research by Frithjof VOSS (1974). Voss has been studying vulcanism, tectonics, the earthquake activity and means of surveyance of active vulcanos since the end of the 1960s.

One must also mention the studies by the cartographer Ulrich FREYTAG, who was employed by the Royal Survey Department in Bangkok from 1973 to 1976 and presented a number of maps and atlases during this time. Important projects were the "Regional Atlas of Asia and the Far East, 1:5.000.000" and the "National Resources Atlas of Thailand".

Similar forms of employment (see W. Röll, 1978, p. 256) were sought by a number of German geographers. Some scientists worked at different South East Asian universities over an extended period of time: G. BURGER in Indonesia, N. STEIN in Malaysia, F. VOSS and D. BRONGER in the Philippines, and W. AUF DER LANDWEHR in Thailand. Additionally, many geographers were employed by international and national institutions or with development projects in the countries of South East Asia: K. HORSTMANN was one of the first of these experts. He worked for the national planning office in Jakarta (Indonesia) from 1955 onwards.

G. ZIMMERMANN (1980) worked on a similar basis in South East Asia and Walter Theodor LORCH visited the Philippines on an assignment of the UNESCO between 1957 and 1962 and published, apart from others, a regional geography of the Philippines (1975). Later on, he lived and worked in Indonesia – also on the assignment of the UNESCO – for a further two years. As experts on German development projects W. Röll, G. Lindauer, U. Scholz and G. Zimmermann travelled to Indonesia in 1969; H. Dürr worked there in the years 1974/75.

Summarizing this information, one can state that German geographers have worked and are still working in all accessible South East Asian countries. Unfortunately, the political situation has hindered geographical research work in Burma, Laos, Cambodia and Vietnam.

4. Australia and New Zealand

The fact that postwar research by German geographers in Australia – apart from some exceptions – started only in the 1970s results from a highly developed Australian geographic science. Australian geographers study all significant research topics and their research essays and books can easily be obtained in Germany. There are also hardly any language problems involved. This is the important difference between Australia and the countries of East and partly of South East Asia.

Nevertheless, not all questions which are of interest to German geographers are researched and answered by Australian geographers. Therefore, it may be quite useful if German scholars visit Australia, in order to analyse geographical subjects, in order to personally observe Australian development processes, in order to learn Australian research methods and also to establish a basis for co-operation between German and Australian colleagues. In the past 30 years, cultural geographical topics have dominated the research of German geographers in Australia. Whereas approximately 15 German cultural geographers have worked in Australia, one can only mention 6 geographers studying physical geography.

As early as the mid 1960s, the geomorphologist Hanna BREMER (1967, 1976) researched the morphology and formation of the arid Central Australia. Previously, Ulrich SCHWEINFURTH (1962) had conducted vegetation geographical studies in Tasmania. At the end of the 1960s, the geomorphologist Ernst LÖFFLER (1969, 1985) followed. He was employed as Principle Research Scientist by the Division of Land Use Research, CSIRO, in Canberra for more than 13 years – only interrupted by his job as Senior Lecturer at the University of Papua New Guinea in Port Moresby in 1975. In the 1970s, Heinz SPÄTH (1983) started his research work on the problem of erosion in the semi-arid Northwestern Australia, and Hartmut VALENTIN (1975) on coastal morphology, which was continued by Ernst REINER (1980).

The cultural geographical research activities in Australia had a comparably wider range and intensity. German geographers especially engaged in *agro-geographical subjects.* At an early stage, i.e. in the mid 1950s, Konrad FRENZEL (1957) conducted studies on agriculture, chiefly on problems concerning artificial irrigation. Jürgen DAHLKE (1973, 1976), who unfortunately died very early, inquired into questions of agriculture and showed regional geographical interrelations. Following a long research visit in 1968, he presented a number of publications in the 1970s, particularly one on the Wheat Belt of Western Australia. In 1975 he wrote a regional geography of Australia. Of far-reaching importance are also his studies on the influence by social groups on the development of Western Australia.

Bruno FAUTZ (1970, 1976) studied physiogeographic factors of agriculture in detail. Dieter JASCHKE (1980) also emphazised this influence factor and stressed the climatic influence. Harald UHLIG (1973) researched rice growing, Stefan von GNIELINSKI (1976) studied wine growing, Klaus ROTHER (1984) investigated the growing of special crops under the influence of South European immigrants. Claudia ERDMANN (1984) conducted studies on German settlements in Southern Australia in 1983. Heinrich LAMPING (1981, 1985) worked on a variety of subjects in the 1980s – on irrigational problems, on agriculture and its influence on soil degeneration as well as on the significance of national measures concerning the development function of agriculture in South Western Australia.

On the whole, the problem of *land development* is often the main topic of geographic analyses. K. FRENZEL (1957), W. SCHMID (decentralization, 1975), Jürgen DAHLKE (mining, 1974), Ernst REINER (Snowy Mountain Project for energy production, 1969) and Heinrich LAMPING (1985) studied the basic questions of Australian land development or analysed the importance of specific elements in this process. Peter SEDLACEK (aluminium industry, 1975) and Lutz FEHLING (iron-ore, 1977) researched individual industries. Meanwhile, Rheinhold GROTZ (1982) conducted a profound study on the entire industrial development in rural South Eastern Australia and analysed industry under the aspect of urbanization.

Town structures and development processes were investigated by many German geographers: Dieter JASCHKE 1975 worked on Darwin, Ernst LÖFFLER 1973 on Canberra, Hans-Werner WEHLING (1975) on Sydney, Heinrich LAMPING (1981) on mining towns. Burkhard HOFMEISTER (1985) analysed the basic structures, the formation and the regional differentiation of Australian towns on a number of research visits.

All in all, it becomes evident that German research activities in Australia present a heterogeneous picture, whereby neither a regional nor a sectoral consequence in the sense of completeness has been achieved. This situation is not surprising on reconsidering the above mentioned fact, that German research work in Australia must be seen against the background of an extensive and manifold, autochthonous Australian geographic science.

Geographic literature on *New Zealand* has only led to a limited number of research tours by German geographers since the 1950s. Their studies, nevertheless, are usually restricted to a single stay. Apart form others, one may mention Walter SCHURIG (1956) with his urban geographical, Erika SELLENBERG (1960) with her agro-geographical, Paul WOLDSTEDT (1962) with his glacial-morphological and Klaus HEINE (1973) with his geo-ecological studies.

Only very few German geographers have conducted long-termed research series in New Zealand. First, one has to mention Ulrich SCHWEINFURTH

(1962, 1966, 1980) who engaged in important regionally dispersed studies on vegetation geography and on soil and erosion problems from the beginning of the 1960s over a period of 20 years. A further physio-geographic subject – the glaciation of the New Zealand Alps – has attracted a number of German geographers: apart from the already mentioned P. Woldstedt, F. JEANNERET (block-glaciers, 1975), Herbert LIEDTKE (pleistocene glaciation, 1980) and Wendelin KLAER (glaciation of the Southern Alps; 1980) researched this topic.

As early as the mid 1960s Wilhelm LUTZ (1981) began studying economic an spatial aspects of New Zealand. He used his latest research tour to inquire into urban geographical matters in 1983. Werner KREISEL (1983) devoted himself to a very special problem. Following the completion of his population and social geographical studies in Hawaii, he researched the difficult situation of immigrants from the South Pacific insular world in New Zealand in 1982.

5. South Pacific

At the beginning of the German geographical postwar research in the South Pacific were geomorphological studies by Alfred WIRTHMANN (1966) in Hawaii in the early 1960s and by Ernst LÖFFLER (1977, 1979) in Papua New Guinea since 1968. Especially Ernst Löffler conducted long and thorough studies on geomorphology and vegetation geography in Papua New Guinea and was Visiting Senior Lecturer at the University of Papua New Guinea. In the mid 1970s Werner KREISEL (1984) worked on settlement and specifically on population and social-geographical subjects in Hawaii.

With the exceptions of Hawaii and New Guinea systematic research work by Germans in the South Pacific insular region was resumed by Hanns J. BUCHHOLZ in 1976. Previously, there were some smaller studies on the changes of the cultural landscape of the islands by Herbert WILHELMY (1971), on the sugar industry in Fiji and Hawaii by Helmut BLUME (1980) and on metropolitan matters of the insular states by Peter SCHÖLLER (1978). Fortunately, there are always some young post-graduates, who spontaneously decide to start research work in foreign cultural regions: e.g. Hanneli HÖLPER (1986) with her studies on tourism in Tonga and Hans Jürgen WIEMER (1985) on questions of land tenureship in Tonga. Hanns J. BUCHHOLZ (1983, 1984, 1986) has meanwhile conducted a number of long research tours in the South Pacific region and has analysed urban and economic development, centralization, the internal national development and the importance of the new Law of the Sea for the small insular states and territories. In 1984 he published the first German comprehensive regional geography of the South Pacific since Walter Geisler's publication in 1939.

Selected Bibliography

I. Western Pacific Region: Pre-War Research

BEHRMANN, W.: Die Insel Neuguinea. Grundzüge ihrer Oberflächengestaltung nach dem gegenwärtigen Stand der Forschung.
Z.G.E. Berlin, Sonderband 1928, p. 191–207.

CREDNER, W.: Siam, das Land der Tai.
Stuttgart 1935.

DOMIN, K.: Beiträge zur Flora und Pflanzengeographie Australiens.
Stuttgart 1928.

FOCHLER-HAUKE, G.: Chinesische Kolonisation und Kolonialpolitik.
Z.G.E. Berlin (1933), p. 108–122.

FRIEDERICI, G.: Wissenschaftliche Ergebnisse einer Forschungsreise nach dem Bismarckarchipel im Jahre 1908.
Berlin 1912. (Mitt. aus den deutschen Schutzgebieten, Erg.-Heft 5).

GEISLER, W.: Australien und Ozeanien.
Wien 1939.

HELBIG, K.: Batavia.
Hamburg 1931.

KOLB, A. Die Philippinen.
Leipzig 1942.

KRÄMER, A.: Die Entstehung und Besiedlung der Koralleninseln.
Stuttgart 1927.

LAUTENSACH, H.: Korea.
Leipzig 1945.

MECKING, L.: Japanische Stadtlandschaften.
In: S. Passarge (ed.): Stadtlandschaften der Erde. Hamburg 1930, p. 109–123.

MECKING, L.: Japans Häfen, ihre Beziehungen zur Landesnatur und Wirtschaft.
Hamburg 1931. (M.G.G. Hamburg 42).

OTTE, F.: China, Wirtschaftspolitische Landeskunde.
Gotha 1927. (P.M., Erg.-Heft 194).

REIN, J. J.: Japan nach Reisen und Studien.
2 vols. Leipzig 1881 and 1886.

RICHTHOFEN, F. v.: China. Ergebnisse eigener Reisen und darauf gegründeter Studien.
Vol. I–V with 2 vols. Atlas. Berlin 1877–1912.

SAPPER, K.: Die deutschen Südseebesitzungen.
G.Z. 21 (1915), p. 624–645.

SCHEIDL, L.: Neuere Beiträge zur Siedlungsgeographie Japans.
P.M. 99 (1955), p. 110–117.

SCHMITTHENNER, H.: Chinesische Landschaften und Städte.
Stuttgart 1925.

II. East Asia

KOLB, A.: Ostasien, China, Japan, Korea. Geographie eines Kulturerdteils.
Heidelberg 1963.

SCHÖLLER, P., H. DÜRR and E. DEGE (ed.): Ostasien.
Frankfurt 1978 (Fischer Länderkunde 1).

II.,1 Japan

AYMANS, G.: Strukturwandlungen und Standortverlagerungen in der japanischen Perlenzucht. Erdk. 19 (1965), p. 112–132.

AYMANS, G.: The Unanimous Society. Remarks on the Generative Behaviour of the Japanese Society in an Extraordinary Year. GeoJournal 4 (1980), p. 215–230.

BOESCH, H.: Japanische Landnutzungsmuster. G. Helv. 14 (1959), p. 22–34.

BOESCH, H.: Zur Bevölkerungsentwicklung in Japan. G. Helv. 32 (1977), p. 45–47.

BOESCH, H.: Wandlungen der japanischen Verkehrsstruktur. G. Helv. 32 (1977), p. 165–168.

ELLENBERG, L.: Shimobashira-Kammeis in Japan. G. Helv. 29 (1974), p. 1–6.

ELLENBERG, L.: Zur Periglazialmorphologie von Ura Nippon, der schneereichen Seite Japans. G. Helv. 31 (1976), p. 139–151.

ELLENBERG, L. AND K. HIRAKAWA: Die Packeisküste Japans. Eiszeitalter und Gegenwart 32 (1982), p. 1–12.

FICK, K. E.: Wurzeln japanischen Wirtschaftsverhaltens. G. R. 36 (1984), p. 114–119.

FICK, K. E.: Japans Gartenanlagen als Raum-, Kultur- und Gesellschaftsphänomen. Praxis Geogr. 14 (1984), p. 26–29.

FLÜCHTER, W.: Neulandgewinnung und Industrieansiedlung vor den japanischen Küsten. Funktionen, Strukturen und Auswirkungen der Aufschüttungsgebiete (umetatechi). Paderborn 1975. (Bochumer Geogr. Arb. 21).

FLÜCHTER, W.: Stadtplanung in Japan. Problemhintergrund, gegenwärtiger Stand, kritische Bewertung. Hamburg 1978. (Mitt. des Inst. für Asienkunde 97).

FLÜCHTER, W.: Central Place Preference by Social Geographic Groups – with Reference to the Tributary Area of Shibuya, Southeast Tokyo. GeoJournal 4 (1980), p. 247–259.

FLÜCHTER, W.: Umweltproblematik und Umweltschutzpolitik in Japan. Grundzüge und Besonderheiten. G.R. 36 (1984), p. 100–113.

FLÜCHTER, W.: Die Bucht von Tokyo. Neulandausbau, Strukturwandel, Raumordnungsprobleme. Wiesbaden 1985. (Schriften des Inst. für Asienkunde 46).

HILLENBRAND, H.: Okinawa – Japans südliche Inselpräfektur. G.R. 32 (1980), p. 413–414.

HILLENBRAND, H.: Zur Entwicklung der Stadt Hiroshima. Würzburger Geogr. Arb. 60 (1983), p. 367–378.

MAYR, A.: Der binnenländische Luftverkehr Japans im Wettbewerb mit anderen Verkehrsträgern. G.R. 36 (1984), p. 126–136.

SCHÖLLER, P.: Centre Shifting and Centre Mobility in Japanese Cities. Proceedings of the IGU Symposium in Urban Geography. Lund 1960, p. 577–593. (Lund Studies in Geography, ser. B, no. 24).

SCHÖLLER, P.: Wachstum und Wandlung japanischer Stadtregionen. Die Erde 93 (1962), p. 202–234.

SCHÖLLER, P.: Ein Jahrhundert Stadtentwicklung in Japan.
Coll. Geogr. 10 (1969), p. 13–57.

SCHÖLLER, P.: Wanderungszentralität und Wanderungsfolgen in Japan.
Erdk. 27 (1973), p. 290–298.

SCHÖLLER, P.: Unterirdischer Zentrenausbau in japanischen Städten.
Erdk. 30 (1976), p. 108–125.

SCHÖLLER, P.: Japan.
In: P. Schöller, H. Dürr and E. Dege (ed.): Ostasien. Frankfurt 1978, p. 325–440. (Fischer Länderkunde 1).

SCHÖLLER, P.: Centrality and Urban Life in Japan.
GeoJournal 4 (1980), p. 199–204.

SCHÖLLER, P.: Technopolis. Ein Zukunftskonzept japanischer Stadt- und Wirtschaftsplanung.
G.R. 36 (1984), p. 94–98.

SCHÖLLER, P.: Die Zentren der Neuen Religionen.
Erdk. 38 (1984), p. 288–301.

SCHWIND, M.: Die regionale Verteilung der japanischen Industrie.
P.M. 98 (1954), p. 280–288.

SCHWIND, M.: Tenri – Stadt des Glaubens.
In: M. Schwind (ed.): Religionsgeographie. Darmstadt 1975, p. 243–247. (Wege der Forschung 397).

SCHWIND, M.: Das Japanische Inselreich.
Vol. 1: Die Naturlandschaft. Vol. 2: Kulturlandschaft, Wirtschaftsmacht auf engem Raum. Berlin/New York 1967, 1981.

II.,2 Taiwan

KOCH, W.: Funktionale Strukturwandlungen in Taiwan. Das Beispiel Lichon im Umland der Millionenstadt Taipei.
Köln 1971. (Kölner Geogr. Arb. 26).

TIMMERMANN, O. F.: Strukturwandel in Taiwan. Europäisierungssymptome im Kulturerdteil China.
Verh. 34. D.G. Heidelberg 1963. Wiesbaden 1965, p. 229–246.

WERNEBURG, K.-H.: Die Handelshäfen Taiwans.
Diss. Bochum 1980.

II.,3 People's Republic of China

DÜRR, H. AND U. WIDMER: Provinzstatistik der Volksrepublik China.
Hamburg 1983. (Mitt. des Inst. für Asienkunde 131).

DÜRR, H. AND U. WIDMER: Ansätze zur Erfassung und Interpretation inter- und intraregionaler Disparitäten in ländlichen Gebieten der VR China.
G.Z. 72 (1984), p. 99–112.

GELLERT, J.: Der Tropenkarst in Süd-China im Rahmen der Gebirgsformung des Landes.
Verh. 33. D.G. Köln 1961. Wiesbaden 1962, p. 376–384.

HEBERER, T.: Nationalitätenpolitik und Entwicklungspolitik in den Gebieten nationaler Minderheiten in China.
Bremen 1984. (Bremer Beitr. zur Geogr. und Raumplanung 9).

HEBERER, T.: Probleme chinesischer Regionalentwicklung.
G.R. 38 (1986), p. 130–136.

KÜCHLER, J.: Stadterneuerung in der VR China.
In: Redaktionskollektiv (ed.): Stadtentwicklungsprozess – Stadtentwicklungschancen: Planung in Berlin, Bologna und in der VR China. Göttingen 1976, p. 137–218.

KUHLE, M.: Permafrost and periglacial indicators in the Tibetan Plateau from the Himalaya Mountains in the south to the Quilian Shan in the north (28°40°).
Z. f. Geomorph. 29 (1985), p. 183–192.

SCHULZ, A.: Die Entwicklung der Stadt Xian, Provinz Shanxi/China, seit den Reisen Ferdinand von Richthofens.
Die Erde 114 (1983), p. 147–164.

STADELBAUER, J.: Gezügelte Bevölkerungszunahme und bleibende Ernährungsprobleme. Dargestellt am Beispiel der VR China.
G.R. 36 (1984), p. 565–572.

TAUBMANN, W.: Problems of Urban Housing in China.
In: Ch.-K. Leung and J. Ch. Chai: Development and Distribution in China. Hongkong 1985, p. 173–212. (Centre of Asian Studies, Occasional Papers and Monographs 61).

TAUBMANN, W.: Stadtentwicklung in der VR China.
G.R. 38 (1986), p. 114–123.

WEIN, N.: Die heutige chinesische Landwirtschaft – dargestellt am Beispiel des Wei-Hi-Beckens.
Z. f. Agrargeogr. 3 (1985), p. 181–200.

WIDMER, U.: Die chinesische Lösshochebene. Ernährungswirtschaft und Ökologie im Widerstreit.
G.R. 33 (1981), p. 88–96.

WIDMER, U.: Neue Strukturen im ländlichen China.
G.R. 38 (1986), p. 138–145.

II.,4 Korea

DEGE, E.: Ein multivarianter Ansatz zur Regionalisierung von Landnutzungsstrukturen – dargestellt am Agrarraum Südkorea.
Kieler Geogr. Schriften 50 (1979), p. 165–182.

DEGE, E.: Industrialisierung und demographische Entwicklung – Raumzeitliche Veränderung der Bevölkerungsstruktur Südkoreas 1960–1980.
Coll. Geogr. 18 (1985), p. 83–107.

II.,5 Hong Kong and Macau

BUCHHOLZ, H. J.: Die Wohn- und Siedlungskonzentration in Hong Kong als Beispiel einer extremen städtischen Verdichtung.
Erdk. 27 (1973), p. 279–290.

BUCHHOLZ, H. J.: Bevölkerungsmobilität und Wohnverhalten im sozialgeographischen Gefüge Hong Kongs.
Paderborn 1978. (Bochumer Geogr. Arb., Sonderheft 10).

BUCHHOLZ, H. J.: Hong Kong. Europäischer Handelsvorposten – Industriekolonie – Transferzentrum für China.
G.R. 38 (1986), p. 510–516.

BUCHHOLZ, H. J. AND P. SCHÖLLER: Hong Kong. Finanz- und Wirtschaftsmetropole. Entwicklungspol für Chinas Wandel.
Braunschweig 1985.

FRIEDRICH, K. AND R. HELMSTÄDT: Deutsche in Hong Kong – Wohnverhalten zwischen Tradition und Transformation.
Z. f. Wirtschaftsgeogr. 29 (1985), p. 38–51.

GOLGER, O. G.: Wohnverhältnisse und Sanierungsmassnahmen.
Bauwelt 42 (1972), p. 1588–1604.

KÜCHLER, J. AND SUM KONG-SUT: Das räumliche Ungleichgewicht Hong Kongs.
Die Erde 102 (1971), p. 141–179.

SCHÖLLER, P.: Hong Kong – Weltstadt und drittes China.
G.Z. 55 (1967), p. 110–141.

SCHÖLLER, P.: Macau – Funktionsumkehr am Überseetor Ostasiens.
In: F. Tichy and J. Schneider (ed.): Stadtstrukturen an alten Handelswegen im Funktionswandel bis zur Gegenwart. Neustadt a. d. Aisch 1984, p. 89–100. (Schriftenreihe des Zentralinstituts für Fränkische Landeskunde und Allgemeine Regionalkunde an der Univ. Erlangen/Nürnberg 25).

SIMKO, H. D.: Strassenverkäufer und die Versorgung der Arbeiterfamilien von Kowloon (Hong Kong) im Umfeld der staatlichen Planungspolitik.
Basel 1983. (Baseler Beitr. zur Geogr. 29).

III. Southeast Asia

KOLB, A.: Südostasien und die Westpazifische Region.
G.Z., Beihefte, Erdkundl. Wissen 58 (1982), p. 146–156.

KÜHNE, D.: Südostasien und die Industrienationen im Lichte der politischen Geographie.
Internationales Asienforum 12 (1981), p. 371–397.

RÖLL, W.: German Geographical Research in Southeast Asia since the Second World War.
Materialien zum Internationalen Kulturaustausch, vol. 6, 1978, p. 253–267.

RÖLL, W., U. SCHOLZ AND H. UHLIG (ed.): Wandel bäuerlicher Lebensformen in Südostasien.
Giessen 1980. (Giessener Geogr. Schriften 48).

UHLIG, H. (ed.): Südostasien – Austral-pazifischer Raum.
Frankfurt 1975. (Fischer Länderkunde 3).

UHLIG, H.: Der Anbau an den Höhengrenzen der Gebirge Süd- und Südostasiens.
Arb. aus dem Geogr. Inst. d. Univ. des Saarlandes 29 (1980), p. 279–310.

UHLIG, H.: Man and tropical Karst in Southeast Asia.
GeoJournal 4 (1980), p. 1–44.

UHLIG, H.: Reisbausysteme und -ökotope in Südostasien.
Erdk. 37 (1983), p. 269–282.

WILHELMY, H.: Reisanbau und Nahrungsspielraum in Südostasien.
Kiel 1975.

III.,1 Philippines

BRONGER, D.: Metropolization: Its Impact to Regional Development. Planning in Developing Countries. The Philippine Case.
Philippine Geographical Journal 26 (1983), p. 113–136.

HAUSHERR, K.: Agricultural Colonization in the Kapatagan Basin, Lanao del Norte, Mindanao, Philippines.
Jahrbuch des Südasien-Inst. d. Univ. Heidelberg 3 (1968/69), p. 100–116.

KOLB, A.: Die Chinesen, Japaner und Inder auf den Philippinen.
Wiesbaden 1974. (Schriften des Inst. für Asienkunde 38).

KOLB, A. (in co-operation with J. EHLERS AND F. VOSS): Das Überschwemmungsproblem in Greater Manila.
Hamburg 1975. (M.G.G. Hamburg 63).

KOLB, A.: Die Entwicklungsphasen von Greater Manila. Der Staat und sein Territorium.
In: W. Leupold and W. Rutz (ed.): Beiträge zur raumwirksamen Tätigkeit des Staates. Wiesbaden 1976, p. 143–158.

LORCH, W. T.: Philippinen.
In: H. Uhlig (ed.): Südostasien-Austral-pazifischer Raum. Frankfurt 1975, p. 344–358. (Fischer Länderkunde 3).

VOSS, F.: Geology and Geomorphology of the Sulu Archipelago.
Z. f. Geomorph. 18 (1974), p. 389–406.

III.,2 Thailand and Vietnam

HOHNHOLZ, J. (ed.): Thailand.
Tübingen 1980.

LÖFFLER, E. AND W. P. THOMPSON et. al.: Quarternary geomorphological development of the lower Mun River basin, north east Thailand.
Catena 11 (1984), p. 321–330.

MATZAT, W.: Einige Besonderheiten des Reisanbaus in Nord-Thailand.
Erdwiss. Forschg. 5, Wiesbaden 1973, p. 117–122.

RIETHMÜLLER, R.: Spontaneous land clearing in Thailand. The Khorat Escarpment and the Chonburi Hinterland.
Giessener Geogr. Schriften 58 (1984), p. 119–260.

THOMASIUS, H.: Die Wälder der SR Vietnam.
P.M. 124 (1980), p. 23–46.

UHLIG, H.: Völkerschichtung und Völkerbewegung in den Gebirgen Thailands im Umbruch der modernen Entwicklung.
Innsbrucker Geogr. Studien 5 (1979), p. 265–291.

UIBRIG, H.: Zur agrarischen und forstlichen Bewertung des Parakautschukbaums und zu seiner Bewirtschaftung in der SR Vietnam.
P.M. 129 (1985), p. 181–186.

WAHNSCHAFFT, R.: Zum Entwicklungspotential des Klein(st)-gewerbes: Der informelle Sektor in thailändischen Fremdenverkehrsorten.
Saarbrücken 1984.

III.,3 Malaysia and Brunei

CORVINIUS, F.: Probleme der Energieversorgung eines tropischen Entwicklungslandes. Das Beispiel Malaysia.
Hannover 1984. (Jahrb. G. G. Hannover).

FRANZ, J. C.: Das Sultanat Brunei. Erdölreichtum und Entwicklungsprobleme.
Nürnberg 1980. (Nürnberger Wirtschafts- und Sozialgeogr. Arb. 32).

KOSCHATZKY, K.: Malaysia. Exportorientierte Industrialisierung und Raumentwicklung.
G.R. 38 (1986), p. 495–500.

KÜCHLER, J.: Penang. Kulturlandschaftswandel und ethnisch-soziale Struktur einer Insel Malaysias.
Giessen 1968. (Giessener Geogr. Schriften 13).

KÜHNE, D.: Urbanisation in Malaysia. Analyse eines Prozesses.
Wiesbaden 1976. (Schriften des Inst. für Asienkunde 42).

KÜHNE, D.: Malaysia.
Stuttgart 1980.

KULKE, E.: Hemmnisse und Möglichkeiten der Industrialisierung peripherer Regionen von Entwicklungsländern. Empirische Untersuchungen über Industriebetriebe in Kelantan/West-Malaysia.
Hannover 1986. (Jahrb. G. G. Hannover).

SENFTLEBEN, W.: Neulanderschließung und raumrelevante Strukturverbesserung von Altland als zentrales Problem der Bodenpolitik in West-Malaysia.
Berlin 1971.

III.,4 Singapore

KÜCHLER, J.: Stadterneuerung in Singapore.
G.R. 23 (1971), p. 67–75.

III.,5 Indonesia

BURGER, G.: Agrare Intensivierungsmassnahmen in Mittel-Java und Probleme ihrer Realisierung.
G.R. 27 (1975), p. 151–161.

BURGER, G.: Das Verbreitungsbild ausgewählter Krankheiten und Ernährungsprobleme in Mittel-Java.
G.Z., Beihefte, Erdkundl. Wissen 43 (1976), p. 98–113.

DRESS, G.: Wirtschafts- und sozialgeographische Aspekte des Tourismus in Entwicklungsländern, dargestellt am Beispiel der Insel Bali in Indonesien.
München 1979. (Wirtschaftswissenschaftliche Forschg. und Entwicklung 36).

DÜRR, H.: Regionalberichte und -pläne für Indonesien.
Erdk. 31 (1977), p. 146–156.

FICHTER, E. M.: Der Einzelhandel im siedlungsräumlichen Gefüge von Djakarta.
Diss. Köln 1972.

GLAESSER, H.-G.: Gegenwärtige Strukturprobleme der Stadt Bandung.
G.Z., Beihefte, Erdkundl. Wissen 43 (1976), p. 114–120.

HORSTMANN, K.: Bevölkerungsverteilung in Indonesien.
Die Erde 95 (1964), p. 167–180.

KÜHNE, D.: New Villages, New Towns and Rural Development in Malaya.
Die Erde 100 (1969), p. 348–358.

LEEMANN, A.: Veränderte Methoden der Reisernte. Ländlicher Wandel in Java.
G. Helv. 33 (1978), p. 173–181.

LEHMANN, H.: Das Antlitz der Stadt in Niederländisch-Indien.
In: H. Louis and W. Panzer (ed.): Festschrift für Norbert Krebs.
Stuttgart 1936, p. 109–139.

METZNER, J. K.: Man and Environment in Eastern Timor: a geoecological analysis of the Bancan-Viqueque Area as a possible basis for regional planning.
Canberra 1977. (Development Studies Centre. Monograph 8).

RÖLL, W.: Indonesien. Entwicklungsprobleme einer tropischen Inselwelt.
Stuttgart 1979.

RÖLL, W.: Siedlung und Agrarstruktur von Pygmäen steinzeitlicher Kulturstufe im zentralen Bergland von Irian Jaya, Indonesien.
Geoökodynamik 2 (1981), p. 79–96.

RÖLL, W. AND A. LEEMANN: Der Ladang-Bau auf Lombok. Studien zum Wandel eines agraren Bodennutzungssystems.
G. Helv. 38 (1983), p. 27–37.

RÖLL, W. AND G. R. ZIMMERMANN: Untersuchungen zur Bevölkerungs-, Siedlungs- und Agrarstruktur im zentralen Bergland von Irian Jaya (West-Neuguinea), Indonesien.
Berlin 1979.

RUTZ, W.: Indonesien. Verkehrserschließung seiner Ausseninseln.
Paderborn 1976. (Bochumer Geogr. Arb. 27).

RUTZ, W.: Javas Bevölkerungsdichte, dargestellt durch seine Bevölkerungsdichtekarte.
Die Erde 108 (1977), p. 115–123.

RUTZ, W.: Die Städte Indonesiens.
Berlin/Stuttgart 1985. (Urbanisierung der Erde 4).

SCHOLZ, U.: Minang Kabau. Die Agrarstruktur in West-Sumatra und Möglichkeiten ihrer Entwicklung.
Giessen 1977. (Giessener Geogr. Schriften 41).

SCHOLZ, U.: Land Reserves in Southern Sumatra/Indonesia and their Potentialities for Agricultural Utilization.
GeoJournal 4 (1980), p. 19–30.

SCHWEINFURTH, U.: Zur Pflanzengeographie der javanischen Bergwelt.
Erdk. 28 (1974), p. 145–148.

STEIN, N.: Die tropische Landwechselwirtschaft. Problematik und moderne Strukturveränderungen, dargestellt am Beispiel Nordsumatras.
G.Z. 60 (1972), p. 322–340.

STEIN, N.: Der Dolok Sinabung. Vertikale Landschaftsgliederung eines Vulkans im nördlichen Batakhochland.
Die Erde 105 (1974), p. 34–61.

ZIMMERMANN, G. R.: Transmigration in Indonesien. Eine Analyse der interinsularen Umsiedlungsaktion zwischen 1905 und 1975.
G.Z. 63 (1975), p. 104–122.

ZIMMERMANN, G. R.: Die Wirtschaftsformationen im südlichen Sumatra (Lampung).
Braunschweig 1980. (Braunschweiger Geogr. Studien 2).

IV. Australia

BREMER, H.: Zur Morphologie von Zentralaustralien.
Heidelberg 1967. (Heidelberger Geogr. Arb. 17).

BREMER, H.: Das australische Nordterritorium im Aufbruch.
G.Z. 64 (1976), p. 13–32.

DAHLKE, J.: Der Weizengürtel in Südwestaustralien.
Wiesbaden 1973. (G.Z., Beihefte, Erdkundl. Wissen 34).

DAHLKE, J.: Die Rolle des Bergbaus bei der Ausweitung des westaustralischen Siedlungsraumes.
G.R. 26 (1974), p. 337–343.

DAHLKE, J.: Das Wirken sozialer Gruppen am Beispiel der Erschliessung des westaustralischen Wirtschaftsraumes.
G. Helv. 31 (1976), p. 179–190.

ERDMANN, C.: Deutsche Siedlungen in Südaustralien.
Erdk. 38 (1984), p. 302–313.

FAUTZ, B.: Agrarräume in den Subtropen und Tropen Australiens.
G.R. 22 (1970), p. 385–391.

FAUTZ, B.: Junge Kulturlandschaftsveränderungen in Queensland.
G.Z. 64 (1976), p. 33–45.

FEHLING, L.: Die Eisenerzwirtschaft Australiens.
Köln 1977. (Kölner Forschg. zur Wirtschafts- und Sozialgeogr. 24).

FRENZEL, K.: Künstliche Bewässerung in Australien.
Bremen-Horn 1957. (Forschungs- und Sitzungsberichte der Akademie für Raumforschung und Landesplanung, vol. 5, Lief. 5).

GNIELINSKI, S. v.: Der Weinbau Australiens. Ein Beitrag deutscher Siedler zur Entwicklung der australischen Kulturlandschaft.
M.G.G. München 61 (1976), p. 205–219.

GROTZ, R.: Industrialisierung und Stadtentwicklung im ländlichen Südost-Australien.
Stuttgart 1982 (Stuttgarter Geogr. Studien 98).

HOFMEISTER, B.: Die strukturelle Entwicklung der australischen Stadt.
G.R. 37 (1985), p. 36–42.

JASCHKE, D.: Die australische Landwirtschaft zwischen klimatischer und marktwirtschaftlicher Herausforderung.
Erdk. 34 (1980), p. 269–280.

LAMPING, H.: Entwicklungsperspektiven der Bergbaustädte im Hamersley-Gebirge Westaustraliens.
Frankfurter Wirtschafts- und Sozialgeogr. Schriften 36 (1981), p. 233–268.

LAPMING, H.: Landwirtschaft und Bodendegradation in Australien.
Frankfurter Wirtschafts- und Sozialgeogr. Schriften 47 (1985), p. 219–238.

LAMPING, H.: Australien.
Stuttgartg 1985. (Klett-Länderprofile).

LÖFFLER, E.: Relief and landform map of Australia.
Canberra 1969.

LÖFFLER, E.: Naturräumliche Faktoren und Landnutzungspotentiale Australiens.
G.R. 37 (1985), p. 4–11.

REINER, E.: Das Snowy Mountain Projekt in Südost-Australien.
G.R. 21 (1969), p. 315–325.

REINER, E.: Zur Küstenmorphologie Australiens.
Berliner Geogr. Studien 7 (1980), p. 201–216.

ROTHER, K.: Der Sonderkulturbau in Süd-West-Australien und seine südeuropäische Trägerschaft.
Erdk. 38 (1984), p. 45–54.

SCHWEINFURTH, U.: Studien zur Pflanzengeographie von Tasmanien.
Bonn 1962. (Bonner Geogr. Abh. 31).

SEDLACEK, P.: Die Standortentwicklung der australischen Aluminiumindustrie.
Die Erde 106 (1975), p. 193–200.

SPÄTH, H.: Flächenbildung in Nordwest-Australien.
Geoökodynamik 4 (1983), p. 191–208.

UHLIG, H.: Die Reisbaugebiete Australiens.
G.Z., Beihefte, Erdkundl. Wissen 33 (1973), p. 411–427.

VALENTIN, H.: Untersuchungen zur Morphodynamik tropisch-subtropischer Küsten.
Würzburger Geogr. Arb. 43 (1975), p. 9–24.

WAGNER, E.: Das Stadtgebiet von Carnavon (Westaustralien) und die Bewässerungswirtschaft am unteren Gascoyne.
Z. f. Agrargeogr. 3 (1985), p. 242–258.

WEHLING, H.-W.: Funktionalbereiche im Grossraum Sydney.
Die Erde 106 (1975), p. 90–105.

V. *New Zealand*

FAUTZ, B.: Die Entwicklung neuseeländischer Kulturlandschaften, untersucht in vier ausgewählten Farmregionen.
Saarbrücken 1970. (Arb. aus dem Geogr. Inst. d. Univ. des Saarlandes, Sonderband 2).

GUBLER, R.: Tauranga – zweitgrößter Exporthafen Neuseelands.
G. Helv. 22 (1967), p. 36–42.

HEINE, K.: Höhenstufung von Zentral-Otago (Neuseeland) – Geoökologischer Überblick.
In: C. Troll and W. Lauer (ed).: Geoökologische Beziehungen zwischen der temperierten Zone der Südhalbkugel und den Tropengebirgen.
Erdwiss. Forschg. 11, Wiesbaden 1978, p. 537–560.

JEANNERET, F.: Blockgletscher in den Südalpen Neuseelands.
Z. f. Geomorph. 19 (1975), p. 83–94.

KLAER, W.: Über den Stand der Vergletscherung in den südlichen neuseeländischen Alpen.
G.Z., Beihefte, Erdkundl. Wissen 59 (1982), p. 25–38.

KREISEL, W.: Gastarbeiter in Neuseeland – Die Pazifik-Insulaner.
Erdk. 37 (1983), p. 303–310.

LIEDTKE, H.: Die pleistozänen Vergletscherungen am Lake Tekapo im Mackenzie Becken.
Berliner Geogr. Studien 7 (1980), p. 323–338.

LUTZ, W.: Der Wirtschaftsraum des High Country (Südinsel Neuseelands).
Frankfurter Wirtschafts- und Sozialgeogr. Schriften 36 (1981), p. 269–330.

SCHURIG, W.: Kaikohe und seine Umwelt. Das Entwicklungsbild einer neuseeländischen Stadt im Nordland.
M.G.G. und des Naturhistorischen Museums in Lübeck 47 (1957), p. 37–132.

SCHWEINFURTH, U.: Stewart Island (Neuseeland).
Die Erde 93 (1962), p. 279–306.

SCHWEINFURTH, U.: Neuseeland. Beobachtungen und Studien zur Pflanzengeographie und Ökologie der antipodischen Inselgruppe.
Bonn 1966. (Bonner Geogr. Abh. 36).

SCHWEINFURTH, U.: Die neuseeländischen Inseln in dreidimensionaler Sicht.
Arb. aus dem Geogr. Inst. d. Univ. des Saarlandes 29 (1980), p. 251–261.

SELLENBERG, E.: Die Entwicklung der Viehwirtschaft in Neuseeland.
Erdk. 14 (1960), p. 115–133.

WOLDSTEDT, P.: Die Vergletscherung Neuseelands und die Frage ihrer Gleichzeitigkeit mit den europäischen Vereisungen.
Eiszeitalter und Gegenwart 12 (1962), P. 12–24.

VI. *South Pacific*

BLUME, H.: Die Zuckerwirtschaft von Fiji und Hawaii.
G.Z. 68 (1980), p. 284–304.

BUCHHOLZ, H. J.: The Role of Small Cities in Spatial Development in South Pacific Island Countries.
Malaysian Journal of Tropical Geography 8 (1983), p. 1–9.

BUCHHOLZ, H. J. (ed.): Australien – Neuseeland – Südpazifik.
Frankfurt 1984. (Fischer Länderkunde 10).

BUCHHOLZ, H. J.: Seerechtszonen im Pazifischen Ozean. Australien, Neuseeland – Ost- und Südostasien, Südpazifik.
Hamburg 1984. (Mitt. des Inst. für Asienkunde 137).

BUCHHOLZ, H. J.: Stadt und Verstädterung im Südpazifischen Raum.
Erdk. 38 (1984), p. 136–148.
BUCHHOLZ, H. J.: Südsee-Inseln. Kleinheit als Entwicklungsproblem.
G.R. 39 (1987), p. 14–21.
HÖLPER, H.: Tourismus in Tonga.
Aachen 1986 (Aachener Geogr. Arb. 18).
KREISEL, W.: Die ethnischen Gruppen der Hawaii-Inseln.
Wiesbaden 1984. (G.Z. Beihefte, Erdkundl. Wissen 68).
LÖFFLER, E.: Geomorphology of Papua New Guinea.
Canberra 1977.
LÖFFLER, E.: Papua New Guinea.
Richmond 1979.
LÖFFLER, E.: Landforms. Papua New Guinea Atlas. A nation in transition.
Port Moresby, o.J.
SCHÖLLER, P.: Rezente Staatenbildung und Hauptstadt-Zentralität im Süd-Pazifik.
Erdk. 32 (1978), p. 228–249.
WIEMER, H.-J.: Agrarstruktur in Tonga.
Göttingen 1985. (Sozialökonomische Schriften zur ruralen Entwicklung 61).
WILHELMY, H.: Wandlungen der Kulturlandschaft in der Südsee.
In: G. Borchert, G. Oberbeck and G. Sandner (ed.): Wirtschafts- und Kulturräume der aussereuropäischen Welt.
Hamburger Geogr. Studien 24 (1971), p. 295–310. (Festschrift für Albert Kolb).
WIRTHMANN, A.: Die Landschaften der Hawaii-Inseln.
Würzburg 1966. (Würzburger Geogr. Arb. 19).

Field research of German geographers in Latin America

by

Jürgen Bähr (Kiel) and Erdmann Gormsen (Mainz)

During the past three decades, research activities of German geographers in Latin America have increased to such an extent that it is hardly possible to overlook the existing literature in its regional as well as its thematic diversity. Therefore the following paper does not try to present a complete account of all the fieldwork and publications of German geographers (see E. GORMSEN and K. LENZ 1987 with comprehensive bibliographies). It rather attempts to present some lines of development which are still effective and visible today, the origins of which, however, date back to the period before the Second World War. It also attempts to point out the thematic emphasis of present day research and to discuss its most relevant results as well as to indicate the still existing gaps in some sections of geographical research. According to the personal interests and research orientation of the authors this is done mainly from the viewpoints of human and social geography. Research papers into physical geography are only considered in this review if they deal with environmental problems related to man. The same applies to the selection of papers in the list of references.

The tradition of geographical research in Latin America

German geographical research in Latin America can look back onto a long standing tradition of scientific investigations. Since the work of Alexander von Humboldt German geographers have made important contributions. Their beginnings were characterized by great expeditions undertaken by natural scientists who presented their findings in extensive books of travel, and who thus gradually reduced the "blanc spaces" on the map of Latin America. Later on the focus of geographic research centred more on regional geographies of certain territories and on investigations of particular topics. The immigration of European settlers who, since the middle of the 19th century filled the "empty spaces" of the subtropical areas of Argentina, Brazil, and Chile particularly stimulated the research into these regions. From the beginning, however, not only geography but other earth sciences, as well as social and historical sciences took part in these studies, above all those disciplines, which are referred to as "Americanistics" (see W. LAUER 1976a).

The substantial reports of the great expeditions combined a variety of field observations with the interpretation of historical sources and an analysis of statistical data. The description of personal experiences and the exotic nature of the areas visited often made these reports to exciting literature for the educated class. Some researchers investigated more specialized topics such as plants and animals, geology and minerals, ethnology and economic development or they searched for areas of potential settlement for German migrants (W. SIEVERS 1893).

Towards the end of the 19th century a more systematic method of regional description was established. This followed the example given by A. HETTNER (1892) in his study on "The Cordillera of Bogotá". This book differs from Hettner's previous rather personal account "Reisen in den columbianischen Anden" (travels in the Colombian Andes) by its concise and factual structure which already gives an indication of his concept of "Länderkundliches Schema" (regional description) which he fully developed later on. Hereby Hettner combines scientific facts and observations about the activity of man to a regional synthesis.

Other scientists who had specialized into a region or country dealt also with and published on a broad spectrum of topics which included aspects of both, physiographical and human geography (c.f. extensive literature reviews of W. SIEVERS 1913, O. QUELLE 1926 and O. BERNINGER 1939/40). Very often the main aim of their work was a regional geography. Important contributions are from K. Sapper, one of the pioneers of tropical geography who over decades published numerous works on the geography of Middle America, from F. Termer, a geographer and ethnologist who wrote a comprehensive regional account on Guatemala and Yucatán and from W. Sievers who already at the turn of the century attempted to compile a regional geography of the whole continent. L. Waibel, O. Schmieder, and C. Troll have continued this work into the post Second World War period and have given new impulses (see below). Excellent reviews on the state of knowledge around 1930 are the contributions by O. Maull, F. Kühn, W. Knoche, C. Troll, and by F. Termer and H. Hagen in the volumes South America and North America respectively of the "Handbuch der geographischen Wissenschaft" (Handbook of the Geographic Sciences) edited by F. KLUTE. At about the same time between 1932 and 1934, O. Schmieder published three regional geographies on Southern, Middle and Northern America in which he first applied his concept of an historical evolution to an entire continent and its larger regions.

Research between the two World Wars and its after-effects

It is largely due to a few personalities who had already worked in Latin America in the twenties and thirties that German geographical research in Latin America regained international recognition after the Second World War and was able to continue its long standing traditions thus maintaining a high degree of continuity. These personalities not only developed and refined their concepts but they also encouraged staff members and students to concern themselves with this continent and thus stimulated further research.

Carl Troll played a most decisive role in the revival of research activities in Latin America after the Second World War (see W. LAUER 1976b with bibliography of C. Troll). His field investigations largely date back to the years 1926 to 1929 when he stayed in Bolivia and undertook, under most difficult logistic conditions, a number of expeditions also to northern Chile, southern Peru, the coastal lowlands of Ecuador, the lowlands and the northwestern regions of Colombia and southern Panamá. Here he gained his fundamental ideas on the classification of climates, on landscape ecology and on the comparative geography of high mountains. During the war years he combined these observations to a comprehensive world wide concept which was assisted by the fact that in the thirties he was also able to visit the mountains of eastern Africa and the Himalayas on larger expecitions.

However it was Troll's work in the Bolivian Andes that proved to be decisive for his later work and this is due to no small measure to the close proximity in which we find here all tropical climates as well as the whole spectrum of tropical vegetation from rain forest to desert and from lowland to high mountains. C. Troll was, however, not only interested in the physical effect of the climate on the natural environment (i.e. periglacial solifluction and patterned ground, delineation of snowline, glacial limits, timber line) but he also studied the impact on man and his role in changing the face of the earth. Hereby he made a substantial contribution to our understanding of the settlement expansion and of the economic exploitation of the high mountains. Troll thus recognized that the most important basis for the development of the ancient American civilisations was their understanding of the ecology of the seasonally humid high Andes (C. TROLL 1943). After the war C. Troll visited the American continent on three shorter travels in connection with congresses in Mexico (1953), Brazil (1956) and with the Humboldt festivities in 1959.

Because of his activities in numerous international associations and in particular due to his presidency of the International Geographical Union from 1960 to 1964 C. Troll was able to reestablish the international reputation of German geography and to make known its achievements in Latin America. His students were able to build on the foundations laid by him. Some of them choose Latin America as regional focus for their research and they established important "modules" for a comparative geography of the world as envisaged by Troll (K. H. Paffen, F. Monheim, W. Weischet, W. Lauer).

Oskar Schmieder began his research work even before C. Troll's involvement in Latin America (cf. W. LAUER 1961 and J. BÄHR 1981 with bibliography of O. Schmieder). Unlike other geographers of his generation he spent a considerable time of his life overseas; for thirteen years he lectured at various universities of South and North America. Between 1919 and 1925 he held the chair for mineralogy at the University of Córdoba and from there he undertook numerous excursions with his students and extensive research trips into nearly all parts of Argentina and into the south of Bolivia. In 1925 Schmieder moved to Berkeley, California where he was a visiting and later associate professor. From Berkeley he visited and investigated Indian settlements in Mexico. Also in Berkeley Schmieder began to develop – in co-operation with C. O. Sauer and other American colleagues – his concept of a historical evolution in the analysis of human geography on which he based his regional geographies (O. SCHMIEDER 1932–1934).

Although O. Schmieder's American studies had found a first completion with these publications the rapid change of the Latin-American environment soon tempted him to study some particular problems. Despite of the world economic depression and the restrictions on travels and research imposed by the Nazi regime he managed, in 1936/37, to investigate together with his then assistant H. Wilhelmy the influence of German colonists on the agricultural opening up of the South-American grasslands (O. SCHMIEDER and H. WILHELMY 1938).

Both these researchers again took up this topic in a more general context during the Second World War and both were able to travel to South America as early as the beginning of the fifties. After his retirement in 1956, O. Schmieder again went to a Latin-American university, the Universidad de Chile, as guest professor. The field studies and travel impressions during this stay in 1958/59 he incorporated into the new edition of his regional geographies. The volume on Middle and Southern America appeared in 1962 (O. SCHMIEDER 1962/63); a Spanish language edition followed in 1965.

From his former students and staff H. Wilhelmy, H. Blume, and W. Lauer in particular have later concentrated on research in Latin America. H. Wilhelmy today can look back on half a century of research work in this area. He visited nearly all countries and landscapes of Middle and South America on his extensive travels. His research interests encompass a wide range of topics extending from Pleistocene geomorphology and vegetation changes in the drylands of northern South America, to questions of tropical transhumance in Colombia and in the Amazonas region as well as to an extensive study on the Maya civilisation (H. WILHELMY 1981) in which he attempts to explain from a geographical point of view the reasons for the rise and fall of the Maya culture (see also the selection of important articles in H. WILHELMY 1980).

Inspite of this great variety in topics there are some major themes in H. Wilhelmy's work; these are questions of colonisation and settlement, a research interest that dates back to his work before the Second World War (O. SCHMIEDER and H. WILHELMY 1938) and above all problems of urban geography which occupied Wilhelmy for decades and which resulted in a "classic"; his book "Südamerika im Spiegel seiner Städte" (South America in the Mirror if Its Cities, H. WILHELMY 1952). This was followed three decades later by a new edition in two volumes (H. WILHELMY and A. BORSDORF 1984/85). In the field of regional geography too, H. Wilhelmy has continued the tradition set by O. Schmieder. This is shown in particular by his great regional geography of the La Plata states (H. WILHELMY and W. ROHMEDER 1963) which, although based on a former draft by Rohmeder on Argentina, was mainly the concept of Wilhelmy.

H. Blume conducted several research trips to the West Indies and besides climatological and geomorphological studies he concerned himself mainly with modern changes of agricultural systems and land tenure (H. BLUME 1961). The results of his investigations were published in a monograph on the West Indies which was also translated into English (H. BLUME 1968). He was able to link his work with older investigations of German geographers in particular with the excellent study of W. CREDNER (1943) who shortly before the outbreak of the Second World War visited the Great Antilles for two months.

W. Lauer was, like O. Schmieder, not only able to visit many parts of Middle and South America; he also spent several years as lecturer and researcher at Latin-American Universities and research institutions. After completion of his Ph. D. thesis under the supervision of C. Troll he went to Kiel in 1951 to take up a position as research assistant to O. Schmieder. He visited Latin America for the first time in 1953 where he spent a year as visiting scientist with the "Instituto Tropical de Investigaciones Científicas" (ITIC) in San Salvador. This interdisciplinary research institution was founded in 1951 through the initiative of the natural philosopher A. Meyer-Abich. Younger scientists from Germany and other countries were to have an opportunity here to work together with local counterparts on projects that were of benefit to the host country. Institutions of a similar nature and similar goals were later established in other Latin-American countries as for instance the "Instituto Colombo-Aleman" in Santa Marta. One of the initiators of this institution was H. Uhlig and a number of his staff worked there for longer periods such as G. Bartels, R. Herrmann, and G. Mertins.

W. Lauer not only undertook research on the climate and vegetation of Central America but also studied the agricultural potential of his host country (W. LAUER 1956). A little later H.-G. Gierloff-Emden also used the opportunity of a stay at the ITIC to study El Salvador and the entire land bridge between the two Americas. W. Lauer returned to Latin America in 1956 and took up a position as Head of Department of Geography at the newly founded Universidad Austral in Valdivia, Chile. Very much in line with Schmieder's concept of historical evolution he investigated the environmental change in the lake area of southern Chile in the post colonial era (W. LAUER 1961). Hereby he was able to make good use of work done by the previous generation of researchers (O. BERNINGER 1929). W. Lauer was succeeded by W. Weischet who spent the years 1959–61 in Valdivia. Weischet conducted a number of research projects concerned with the geomorphology and climatology of Chile, and he also published an extensive regional monograph of Chile (W. WEISCHET 1970).

Leo Waibel, the third great personality of the pre-war period, gained his early overseas experience in Africa. The decisive influence that led to his concept of economic geography however was due to his Mexico travels in 1925. Linking up with research of K. Sapper, carried out during the 1890's he was able to apply empirically his new ideas about economic geography. Thus during this first trip to the New World he developed the new term "Wirtschaftsformation" explaining the regional interdependence of economic and social aspects of agriculture and their relations to the environmental conditions. Apart from this he applied the law of Thünen to geography (G. PFEIFER 1952). L. Waibel's observations and experiences in the American tropics and subtropics were combined impressively into a theoretical framework in his treatment of the economic geography of Mexico. The completion of L. Waibels work in Mexico represented an extensive regional geography of the Sierra Madre de Chiapas (see G. PFEIFER and G. KOHLHEPP 1984 with a bibliography of L. Waibel). As a former student of Hettner he followed his concept of regional geography, however, he expanded on it and tried to incorporate his own ideas and observations on economic features which were based on his field experiences.

It was only after his enforced retirement by the Nazi regime that L. Waibel was able to continue his research work in Latin America. Just before his final emigration to the United States he visited, in 1937, Guatemala and Costa Rica on a private trip. From 1946 to 1950 L. Waibel was a government consultant for regional planning and for various colonisation projects in Brazil and laid the foundation for an applied tropical research. As researcher and teacher L. Waibel had a decisive influence on the Brazilian geography. A whole generation of important Brazilian geographers such as O. Valverde, L. Bernardes, S. Faissol, P. P. Geiger. H. O'Reilly-Sternberg were trained by Waibel and thus became familiar with his methods of geographical field work. This became important for the possibilities of German research in Brazil after the Second World War.

From L. Waibel's students and collaborators (W. Credner, G. Pfeifer, J. Schmithüsen) it was mainly G. Pfeifer who continued the Latin America tradition of his former teacher. Already from 1928 to 1932 when he was with C. O. Sauer at the Department of Geography of the University of California he had his first contacts with the Latin-American culture. In addition to his investigations into the agricultural geography of California he studied, in northwest Mexico, the regional differentiation of agriculture as a consequence of its different historical backgrounds and economic goals. To G. Pfeifer we also owe the posthumous publications of a great deal of the Brazilian research work of L. Waibel who died unexpectedly in 1951 (L. WAIBEL 1955, G. PFEIFER and G. KOHLHEPP (1984).

At the suggestion of L. Waibel, G. Pfeifer carried out a number of research projects in Brazil (see G. PFEIFER 1981) which were later continued by Pfeifer's former students such as G. Kohlhepp and G. Glaser. Former students and staff of G. Pfeifer also worked in other parts of Middle and South America; F. Monheim and A. Kessler in Bolivia and Peru, and F. Tichy and E. Gormsen in Mexico.

In contrast to this, the relatively long immediate postwar involvement of a group of German geographers (F. Machatschek, W. Czajka, G. Fochler-Hauke, K. Hueck) in teaching and research in Tucumán, Argentina did not lead to a comparable follow up. Apart from some important research projects in Patagonia mainly by W. ERIKSEN (1970), the former considerable German influence in the development of geography in Argentina does no longer exist and this country appears to be somewhat neglected which is surprising in view of its size and significance.

Organization and support of research work after the Second World War

While for the time before the Second World War one can nominate clear thematic and regional centres for the research work of German geographers and can also allocate these to the different scientists and their followers, for the post war period this is no longer possible. The last three decades are characterized by an enormous variety of topics and methods as well as a great number of research areas (see E. GORMSEN and K. LENZ 1987). There are several reasons for this: At the beginning of the sixties there was an enormous growth in staff numbers at existing universities and in addition to this a great number of universities were newly established in line with the growth in student numbers. This brought with it a considerable expansion in research activities. Looking at the applications by geographers for research grants that were received by the German Research Foundation (Deutsche Forschungsgemeinschaft, DFG) between 1961 and 1981 one can see that exactly half (878 out of 1772) were for research in overseas countries. Latin America and the Middle East were with 20% each the two areas with the greatest number of applications (E. LICHTENBERGER 1984).

Another reason was the considerable improvement in travel possibilities. The aircraft made access to distant countries easier and the infrastructure in many developing countries was significantly upgraded so that it was no longer necessary to travel by mule or horse as in the twenties. In addition one could much easier obtain research grants. The "Deutsche Forschungsgemeinschaft" (German Research Foundation) which succeeded the "Notgemeinschaft der deutschen Wissenschaft" was founded in 1951 and the "Stiftung Volkswagenwerk" (Volkswagen Foundation) ten years later. Other foundations too have contributed considerably to the growth of empirical field research in Latin America during the last three decades.

Besides large project, some of them of an interdisciplinary character, these foundations always have supported research work of smaller groups and individuals. If one compares the relationship between financial expense and scientific results, some aspects can be mentioned in favour of individual research since, as a rule, motivation and engagement are particularly high here; individual research therefore will maintain its position in the future. A disadvantage however is the fact that problems of an interdisciplinary nature are often being neglected since these go beyond the capacity of an individual researcher or even a small working group. Thus, great part of German geographers working in Latin America do focus on a single country or region and work relatively specialized on a limited range of problems using the appropriate methods.

Larger projects have without doubt the advantage that wide ranging research topics and problems can be tackled. Because of the generally longer period over which a research grant is given, there are better preconditions for an intensive study of the research objects. These larger projects also facilitate the co-operation between researchers of different disciplines and they also tie the research activities much more closely with corresponding work in the host countries. Within German geography larger projects have however remained rather the exception than the rule and this is probably due to the fact that a certain degree of "bureaucratisation" is often connected with these projects. This is shown for instance in the experiences of some special research programmes on Latin America (Sonderforschungsbereiche) which were sponsored by the German Research Foundation at the beginning of the seventies. In some cases, there were endless discussions on the theoretical basis of the research which led to a rather slow implementation of the actual field work (W. LAUER 1976a).

Nevertheless, a number of larger projects were successfull and produced a great deal of valuable results. Amongst these are the "Schwerpunktprogramme" (priority projects) that are financed by the German Research Foundation and are carried out by individual disciplines or groups of disciplines from several universities. There are also interdisciplinary teams financed by other foundations (W. LAUER 1976a). Amongst these, various projects of the *"Arbeitsgemeinschaft Deutsche Lateinamerikaforschung"* = ADLAF (German Association of Latin-America Research) have gained significance.

The ADLAF, founded in 1965, constitutes a number of departments and individual scholars inside and outside the universities that are continually engaged in Latin-American research. Originally it consisted mainly of social scientists (anthropology, political science, economics); but very early, geographers also joined the group. Today the Association encompasses 22 departments and between 150 and 200 researchers amongst them 26 geographers (W. LAUER 1987). According to its statutes the ADLAF endeavours to encourage the exchange of ideas and experiences between the different scientists and disciplines. It organizes annual meetings where topics dealing with Latin America are discussed. The proceedings are usually published (see for instance D. W. BENECKE et al. 1986; G. KOHLHEPP ed. 1987).

Since 1971 the ADLAF has also initiated its own research projects of a multidisciplinary character. In some of these projects geographers played a significant part, as for instance in the project "Development problems in non-tropical Latin America from a historical, geographical and regional policy perspective – a case study from Chile" (Entwicklungsprobleme im aussertropischen Lateinamerika in historischer, geographischer und regionalpolitischer Sicht – Modellstudie Chile). This project was based at the universities of Köln and Bonn. The individual research activities were centred around the problem of urban-rural disparity. Geographers investigated the population dynamics within different areas and tried to explain these by analyzing the socio-economic conditions of the areas of origin and destination (W. LAUER 1976c).

In another research project of the ADLAF that was started in 1974 the economic geography section of the University of Hamburg was involved. The urbanization, industrialization and regional development outside the large agglomerations was investigated taking Central-American examples (for example M. MORALES and G. SANDNER 1982). The research project was supported by the Volkswagen Foundation as is the entire work of the ADLAF.

Further impulses can be expected from the currently running project of this foundation which is called "Fundamental developments in Latin America, Asia and Africa" (Grundlegende Entwicklungen in Lateinamerika, Asien und Afrika). Of the Latin-America related research work within this project two are geographically oriented. These are a project by J. Bähr (University of Kiel) and G. Mertins (Universitiy of Marburg) on "Recent tendencies in the process of urbanization in Latin-American cities" (Jüngere Tendenzen des Urbanisierungsprozesses in lateinamerikanischen Grosstädten) and a project by H. Nuhn (University of Hamburg) on "Polarized settlement development and decentralisation policies in Middle America" (Polarisierte Siedlungsentwicklung und Dezentralisationspolitik in Mittelamerika).

More regionally oriented is the large research project that has been carried out for several years by G. Kohlhepp (University of Tübingen) and his coworkers in the Amazon area of Brazil and the neighbouring states. This project was also funded by the Volkswagen Foundation but also supported occasionally by the DFG. Its aim is the analysis of spatial development strategies for the settlement and economic development of peripheral areas. Special interest is placed on the conflicts that have arisen, as a result of state planning, particularly in the Brazilian part of the Amazon between Indians, small holders, agrobusiness and industrial and energy related large projects (G. KOHLHEPP 1983).

Questions of regional analysis, the assessment of the natural potential, the results of planning implementations and land use conflicts in other parts of Brazil and, more recently, in eastern Paraguay, are also at the centre of research activities of the Tübingen group around Kohlhepp (G. KOHLHEPP 1984). The present research work links up with older activities that were mainly centred around southern Brazil and were concerned with settlement and space related development processes (G. KOHLHEPP 1975).

Of all the research projects in Latin America the DFG funded "*Mexico project*" was without doubt the largest. The field investigations took place between 1962 and 1978 and some of the evaluations are still going on and await publication. Besides the Americanist P. Kirchhoff as main initiator and the ethnologist F. Termer the two geographers G. Pfeifer and W. Lauer have from the beginning contributed substantially to the formation of the project.

The project was interdisciplinary and bilateral and its goal was the investigation of the cultural development of man from the beginnings to the present day within a clearly defined area. The area selected was the highlands of Puebla-Tlaxcala because it was presumed that this area has been settled continuously since man moved into this region and because archaeological and historical sources proved to be favourable.

The major disciplines in the project were archaeology, prehistory and other sections of history, ethnology, and human geography (in particular historical, agricultural, and urban geography). Although the humanities were at the centre of the project natural sciences were included from the beginning since it was clear that many problems regarding the relations between man and the natural environment could only be solved with their collaboration. Within the framework of the Mexico project 19 monographs have so far been published in an own book series "Das Mexiko-Projekt der DFG" (The Mexico project of the German Research Foundation).

A bibliography that appeared in connection with the 1976 International Congress of Americanists in Paris listed some 400 titles. Since then about 200 publications can be added. Amongst the 105 scientists from both countries who participated in the project there were 26 geographers (W. LAUER 1987). Only a few of them can be mentioned here.

F. Tichy was the first who researched the available literature and other documents regarding the study area and together with his cartographic work he set the organisational and geographic basis for the entire project (F. TICHY 1968). Later on he worked in particular on settlement patterns and discovered the influence of cosmological ideas on the orientation of settlement networks in pre-Spanish times (F. TICHY 1976). The investigations of U. EWALD (1976) and H. J. NICKEL (1978) referring to the Mexican hacienda were also historic geographical as were those of K. TYRAKOWSKI (1975) on the development of rural settlements and of W. TRAUTMANN (1983) on the changes in agricultural and settlement patterns after the Conquista. More closely related to the present time are the studies of E. GORMSEN (1971 and 1981) on urban structures and on the central place system under special consideration of periodic markets. Also dealing with the present situation are the studies of H. J. SANDER (1976) on social structures in the rural areas and by E. SEELE (1970) on changes in the agricultural landscape. Amongst the studies in physical geography the work of W. LAUER and his collaborators about the development of climate and vegetation (W. LAUER 1981) and of K. HEINE (1978) on soil erosion and the relationships between man and geomorphological processes are worth mentioning.

Major themes of present research activities

With the brief description of the regional and organisational framework of the Mexico project we have already touched on a central theme in the research of German geographers in Latin America: *the conflict and adjustment between man and environment.* This theme which is multidisciplinary and connects like no other topic both major sections of geography, already fascinated C. Troll in the twenties. The special environment of tropical mountains with its lack of winter and snow, cause the altitudinal belts of vegetation and land use to reach much greater altitudes than in the temperate zones. The change from humid to dry and from cold to warm climates occurs over a very short distance and the great variety of soils and rock types enforces an adaptation of land use by man and a regional differentiation of economies over small areas and thus stimulates the advancement of civilisation in a particular way (C. TROLL 1943).

As in the high altitudes of Peru, Indian civilisations also developed on the Mexican meseta. The region that was investigated by the Mexico project belongs to this meseta. The task of the research that was undertaken in the Puebla-Tlaxcala highland, however, was not primarily directed to assess these excellent achievements. Its aim was rather to investigate the change of the environment that was caused by man and his activities starting from the natural landscape onto which man had not yet set foot at the end of the last ice age to the present time. This meant that the historical development of events (for the most important epochs) had to be complemented by spatial data. Historical and geographical research methods were thus integrated (W. LAUER 1987).

Based on geomorphological, pedological, tephra-stratigraphic, palynological and climatological data it was possible to reconstruct the landscape for various epochs and it was shown that the area around Tlaxcala experienced seven settlement phases between 1600 B.C. and 1521 A. D. The reconstructed settlement and field pattern and the research into the material culture permitted to distinguish clearly the different use of space and the mobility of population from epoch to epoch. However already in earlier periods there were significant changes to the natural environment caused on the one hand by man on the other by climatic changes and related changes in vegetation. This means the history of man has always been closely linked with the history of nature.

However, man has also quite early influenced and often depleted the environment he lived in by expansion of his settlements and land use. One of the most disastrous consequences was the soil erosion which reduced the agricultural possibilities of large areas and led to irreversible losses and even to desertification (K. HEINE 1983).

The man-environment problem is also at the centre of another smaller DFG supported project: "Ökosystem und Mensch im Gebiet der Kallawaya Bergbevölkerung in Bolivien" (Ecosystems and man in the area of the Kallawaya population in Bolivia). Researchers of several disciplines study life and environment of this Indian group as a representative example of the influence of man on a high mountain ecosystem. The land use of this area shows a distinct vertical differentiation and stretches from irrigated fields on terrassed slopes in the lower altitudes to grazing with lamas and alpacas in the higher mountains (W. LAUER 1984).

The inventory of the natural resources, the traditional life style and the social structure of these people showed that natural environment, cultural landscape and agriculture have been closely interlinked for a long time. The close interrelationship between man and environment in the Kallawaya culture is expressed in highly developed forms of adaptation (L. MAHNKE 1982). The basic economic concept has always been to minimize risk and not to maximize gain. Through their agricultural system being spread over three altitudinal zones the Kallawaya people are not only relatively self sufficient in their food production they also have a greatly reduced harvest risk (W. SCHOOP 1982).

Closely linked thematically with the man-environment problem is a second research focus of German geographers in Latin America. This one can be described as "*Land occupation and colonisation*" (Landnahme und Kolonisation) and its aim is the research into the pioneer front in natural or sparcely populated environments. This research focus too can look back over a longer tradition and originates mainly in the work of O. Schmieder and his students (see above). L. Waibel also concerned himself with this topic during his stay in Brazil (L. WAIBEL 1955). While the investigations in the prewar period were largely concentrated in the nontropical areas or the very margins of the tropics (for instance O. SCHMIEDER and H. WILHELMY 1938), from the 1960s they have moved to the tropics because of the colonisation efforts that are taking place there. Today there are five well researched areas which stretch from the far south of the continent to Central America:

1. In the regions outside the core areas of Argentina and Chile the colonisation of settlers dates back to the middle of the last century when the newly independent nations were trying to achieve a greater integration of the peripheral areas of their states. With the rapid increase in population towards the turn of the century this colonisation which was originally motivated by political reasons became an economic and social necessity.

The colonisation in the lake area of southern Chile which was to a large degree carried by German immigrants (since 1848) was investigated by W. GOLTE (1973). W. ERIKSEN (1970) and C. C. LISS (1979) concerned themselves with the agricultural development of eastern Patagonia. This is particularly interesting since the areas along the margin of the Andes experience a new factor of development: tourism. The colonisation of the extreme south of Chile is still not completed today; the steep, deeply dissected, and rain forest covered southern Cordillera offers only few possibilities (J. BÄHR and W. GOLTE in W. LAUER 1976c). With the example of Aisen, A. BORSDORF (1987) has demonstrated the limitations of development in Westpatagonia.

2. The settlement expansion in the Gran Chaco which already arose the interest of O. Schmieder and H. Wilhelmy in the thirties, has later been researched in more detail by J. BÜNSTORF (1976). He tried to establish a model of colonisation as successions in the processes of opening up land for development. In addition he studied the problems of agriculture at its climatic margin. The colonisation of eastern Paraguay has recently been studied by G. KOHLHEPP (1984).

3. Kohlhepp has also studied problems of colonisation in various areas of Brazil as for instance in the pioneer zone of coffee production along the margins of the tropics (G. KOHLHEPP 1975). His particular effort however was directed towards the critical study of land development in the Amazon basin and its socio-economic and ecological consequences. He and his coworkers were able to prove the failure of the development of new areas which was undertaken with great effort and expense as an alternative to land reform. The economic development of these marginal areas has worsened the social problems of the country rather than improved them.

4. Similar projects of land development along the eastern slope of the Andes do not parallel the Brasilian Amazon colonisation in its dimensions. The danger of severe ecological destruction is therefore considerably smaller. However, here too the pioneer front has not improved the standard of living amongst the rural population and has not aided in bringing about a reduction of the population pressure in the settled areas. Systematic research into efforts of colonisation at the margins of the Andes were started by F. Monheim and his students already in the sixties (F. MONHEIM 1965, 1977, W. SCHOOP 1970), regionally concentrating on Bolivia and Peru. The problems of colonisation were linked up with the more general questions of agrarian reform (F. MONHEIM 1981).

At the same time W. BRÜCHER (1968) studied the process of colonisation at the foot of the Colombian Andes and shortly later A. MAASS (1969) and V. JÜLICH (1975) investigated similar problems in Peru. This enabled W. BRÜCHER (1977) to draw some general conclusions. According to him there is a striking contrast between the size of the area under land use and the actual agricultural possibilities. For the entire colonisation area between Colombia in the north and Bolivia in the south he estimates a population of between 1.5 and 2 millions; this corresponds to a mere 3–4 % of the population of the four states. The number of the actual settlers is even considerably lower (cf. C. BORCHERDT and S. SCHAER-GUHL 1985).

5. Like the tropical countries in the Andes most Central American states possess at their eastern side extensive and not yet settled forest areas. For this region too the beginnings and the process of agricultural colonisation have been studied by German geographers. The first studies were untertaken, in Costa Rica, in close co-operation with the "Instituto Tierras y Colonización" (G. SANDNER 1961, G. SANDNER and H. NUHN 1971); later on, the other countries were included into the research project (H. NUHN 1976). The experiences here are similar to those of the regions mentioned before. The colonisation projects that were pushed ahead by the governments were expensive and only included a small number of settlers. They were not able to make any significant change to the negative social and economic conditions created by uncontrolled settlement expansion.

The examples discussed show quite clearly that colonisation can not replace an overall agrarian reform which should also include a change in landed property. This is valid not only for economic reasons but also because of the limited carrying capacity of the tropical soils (for agricultural purposes). However, even a wide ranging redistribution of agricultural land may often not be enough since the land resources are limited as shown by the case studies of K. ROTHER (1977a), G. MERTINS (1979), and F. MONHEIM (1981).

Closely connected with the unsolved problems in the rural areas is the rapid urbanization process, in particular in the capital cities. Parallel to the population growth, the cities not only experience a substantial spatial expansion but also a profound morphological, functional, and social change. Since H. WILHELMY (1952) first studied this topic in the early fifties and gave a review of the state of knowledge at that time, the theme *"Urban pattern and urban development"* (Stadtstruktur und Stadtentwicklung) has remained an important research field for German geographers. Many studies have been carried out, also including towns of medium and small size, using methods of geographical field work, the analysis of maps, aerial photographs, statistics and other data sources such as interviews and questionnaires.

Extensive monographs exist for cities like Barquisimeto in Venezuela (E. GORMSEN 1963), Chochabamba (E. KLAHSEN 1983) and Santa Cruz (G. KÖSTER 1978) in Bolivia, as well as Santiago de Chile (J. BÄHR 1978). In addition to this there are papers on Puebla (E. GORMSEN 1981) and Minatitlán (E. BUCHHOFER 1984) in Mexico, Ciudad Guayana (C. BORCHERDT 1975) and the towns of the Llanos (C. BORCHERDT et al. 1985), Bogotá (W. BRÜCHER 1969), Quito (W. D. SICK 1970), Brasilia (G. PFEIFER 1962,

V. RÖNNICK 1985), Buenos Aires (W. CZAJKA 1959, J. BÄHR 1979b), and Mar del Plata (E. KROSS 1972).

This work provided a good basis for a comparative study of urban areas at different scales – from national to continental level. Such studies on a national level are the investigations of E. GORMSEN (1966) on small towns in the Tlaxcala area, Mexico, the work of W. SCHOOP (1980) on Bolivian Departments centres, and of K. ROTHER (1977b) and A. BORSDORF (1976) on towns in middle and southern Chile respectively. The book of G. SANDNER (1969) on Central American capital cities represents a comparative study encompassing different countries as does the publication of H. NUHN (1981) on the urban system of the same region.

It was again H. Wilhelmy who together with his student A. Borsdorf summarized systematically in two volumes the results of the research on urban geography in Latin America (H. WILHELMY and A. BORSDORF 1984/85). While in the first volume comparative aspects of urban geography dominate, the second volume is more in the form of monographs. The fact that H. Wilhelmy and A. Borsdorf are able to discuss in detail "Models of Latin-American cities" is to no small part due to the availability of relevant publications by other researchers. In 1976 already J. Bähr discussed a structural model of the Latin-American city. Basis for this publication were available empirical investigations on urban patterns in different Latin-American regions and his own research in Santiago. This has finally also led to the revised publication of the model by J. BÄHR and G. MERTINS (1981).

Since the Latin-American large cities are presently subject to rapid changes these models can only serve as "snapshots". In addition the change in structure does not take place simultaneously in all cities and there are remarkable differences between small and large cities and also between cities of different countries. This differentiation can be explained by the differences in natural population growth being often out of phase and differences in migration behaviour. In order to account for these differences the models of A. BORSDORF (1976) and E. GORMSEN (1981) lay more stress on causal relationships.

Abstract concepts of this kind always bear the risk to neglect significant regional differences such as exist between southern and middle America and even more between the former Spanish and Portuguese colonies. Therefore, the models should be understood as working hypotheses for further research and since for many countries of Latin America we have not enough empirical knowledge we can not yet establish whether the generalisations made in the models are always valid.

The discussions about models has proved stimulating and has led to an increase in the study of particular urban-geographical problems. This applies for instance to the changes in the downtown areas (R. MITTENDORF 1984,

P. GANS 1987), the development of residential areas for the middle and upper classes (K. POPP 1985), the supply of housing for the lower social strata (G. MERTINS 1980), and above all for the spatial growth and consolidation of squatter settlements at the city margins (G. MERTINS 1980, F. BUCHHOFER 1982, H. PACHNER 1982, J. BÄHR and G. KLÜCKMANN 1985, R. KRETH 1985).

So far in this paper only three main priorities of German research activities in Latin America have been discussed because these are the ones with the longest tradition. This, however, does not mean that in other fields of geography, German researchers have not made considerable progress and contributed to the solution of particular problems. Only a few of these activities can be mentioned here.

The reviews and list of references on *population geography and geography of tourism and leisure* by J. Bähr and E. Gormsen respectively (in E. GORMSEN and K. LENZ 1986) show that both fields of research have only recently gained greater interest. The same applies to *industrial geography* including the *study of mining*. This is due partly to the longer tradition of research into agricultural geography, partly to the fact that agricultural areas tend to be of particular interest to geographers simply because of their spatial extent even though they may not be the most important areas for the economy as a whole. The small number of investigations about mining and industry include the work of G. KOHLHEPP (1968) about southern Brazil, of H. D. HAAS (1976) about the West Indies, of G. SCHERM (1982) about the Guayana countries, of K. KULINAT (1983) about Venezuela, of W. BRÜCHER (1975) about Colombia, of S. M. UMMENHOFER (1983) about Ecuador, and of W. MIKUS (1985) about Peru.

Apart form investigating singular problems, German geographers have always concerned themselves with *regional synthesis*. After the first continent-wide monograph by W. SIEVERS (1893) and the well conceived regional geography bei O. SCHMIEDER (1932/33 and 1962, see above), it was only G. Sandner who, together with the sociologist H. A. Steger and other co-autors, compiled a regional geography for the whole of Latin America (G. SANDNER and H. A. STEGER 1973). Their approach differs from that of the older publications; they try to combine different points of view to achieve a critical understanding of the political, social, and economic tensions and their spatial and temporal relationships. In this connection, the authors emphasize the regional disparities in the entire subcontinent and its regions. A similar approach is followed by SANDNER (1985) in his book on Central America and the far Caribean West and by H. NUHN (1985) in his collection of papers on "Crisis area Middle America" (Krisengebiet Mittelamerika).

More numerous are the regional geographies of single countries or groups of countries. Because of its size and the breadth of the topics dealt with both in physical and human geography the book by H. WILHELMY and W. ROHMEDER (1963) on the La Plata countries ranks expecially prominent; similarly extensive and complete in its treatments are the books by H. BLUME (1968) on the West Indies, by H. G. GIERLOFF-EMDEN (1970) on Mexico and by W. WEISCHET (1970) on Chile. The regional geographies on Chile by J. BÄHR (1979a) and on Brazil by J. MÜLLER (1984) are more oriented towards social and economic geography.

The position of Latin America research by German geographers

In order to assess the position of German Latin-American research one has to consider three aspects: its role in the framework of the general Latin America research, its position in relation to similar research in other countries, and its significance for the host countries and regions in which the research is conducted. The following notes do not attempt to give a final evaluation of the research results; it is too early to see which achievements have influenced the progress of research most significantly and which will last for a longer period.

Until the Second World War Latin America research consisted of two main lines: there were the traditional *Americanistics* with ethnology, archaeology, and linguistics as well as colonial history and there was *geography* with its subsections in the framework of regional geography. In the last two decades the *social sciences* have entered the field with their studies of present day problems of social change, of the economic development, and of the frequently changing political processes both internally and externally.

Many of the above mentioned research activities as for instance urban and population geography or geography of tourism and leisure or the studies on the agricultural colonisation have concerned themselves with such problems of change. They have thus contributed to the formulation of developmental theories which, for their part, have received significant impulses by the local Latin-American social scientists. Geography and other disciplines are thus linked in thematic research areas, and in the framework of the "German Association of Latin America Research" (ADLAF), there is close co-operation and exchange of ideas among neighbouring disciplines.

One of the main strength of German geography has always been the *empirical field research* which for several reasons has been developed much later by Latin-American scholars. This research leads to the recognition of processes that operate spatially and it therefore represents, like the analysis of statistical data at different scales, an important tool for the spatial differentiation being imperative for the evaluation of regional development. This is important since many misunderstandings about countries and regions are caused by sweeping generalisations. Geography generally remains linked to concrete features in a concrete area and because of this it still persues the aim of a problem oriented regional geography which contributes to the understanding of spatial interrelationships.

The impact of German geographical research in Latin America varies certainly from country to country. This is partly due to the fact that publications in German are, like in many other parts of the world, hardly taken notice of and financial support for translations is generally not available. Another factor is the strong position of North-American geographers. Although there are relatively few who work outside North America those who do concentrate to a great deal on Latin America and thus further strengthen the position of the English language. Their much closer proximity to their research area is of course also a favourable factor. With the "Conference of Latin-American Geographers" USA geographers have set up an association devoted entirely to work in Latin America. They have built up connections to various Latin-American institutions and with annual meetings and a regularly published yearbook they are very effective publically.

The impact of German research, on the other hand, depends on the intensity and continuity of the direct contact with the respective country. To maintain such contacts and connections often proves difficult for the single researcher since he generally is only able to stay in his research area for a few months at the most. Larger projects generally offer better possibilities of maintaining this contact since they usually possess an institutional basis in the host country and/or closely co-operate with local researchers. But here too German research is in a difficult position since it lacks direct governmental support and incorporation into the cultural policy as is for instance the case with French projects.

In addition to these problems there are the great differences in the level and status of local geographical research in the various countries. Geography departments that are well endowed with equipment and staff and which engage in research as well as in teaching are rare. They are to be found mainly in Mexico, Brazil, Chile, and Venezuela. Often the work of the departments is concentrated towards the training of geography teachers and therefore one may find a local counterpart rather in other disciplines of social and earth sciences than in geography. Nevertheless some implementation of geographical research takes place in the framework of regional planning activities either through official co-operation or through personal links with relevant institutions.

Despite some tendencies in many Latin-American countries to reglement more strictly research work untertaken by foreigners there still exists a positive attitude towards single scientists and larger projects. This has often led to long standing co-operation. This certainly justifies the continued support for geographic field research in Latin America and other overseas regions which will remain one of the central tasks of our discipline.

Bibliography

BÄHR, J.: Santiago de Chile. Eine faktorenanalytische Untersuchung zur inneren Differenzierung einer latainamerikanischen Millionenstadt.
Mannheim 1978. (Mannheimer Geogr. Arb. 4).

BÄHR, J.: Chile.
Stuttgart 1979, 2. ed. 1981. (Klett-Länderprofile). [1979a].

BÄHR, J.: Gross Buenos Aires. Zur Bevölkerungsentwicklung der argentinischen Metropole.
Innsbrucker Geogr. Studien 5 (1979), p. 151–172. [1979b].

BÄHR, J.: Oskar Schmieders länderkundliches Lebenswerk (mit Schriftenverzeichnis).
Kieler Geogr. Schriften 52 (1981), p. 1–10.

BÄHR, J. AND G. KLÜCKMANN: Sozialräumliche Differenzierung von Wohngebieten unterer Einkommensgruppen in lateinamerikanischen Metropolen. Die Beispiele Santiago de Chile und Lima.
Ibero-Amerikanisches Archiv, N.F. 3 (1985), p. 283–314.

BÄHR, J. AND G. MERTINS: Idealschema der sozialräumlichen Differenzierung lateinamerikanischer Grosstädte.
G.Z. 69 (1981), p. 1–33.

BENECKE, D. W. et al. (ed.): Desarrollo demográfico, migraciones y urbanización en América Latina.
Regensburg 1986. (Eichstätter Beitr. 17).

BERNINGER, O.: Wald und offenes Land in Südchile seit der spanischen Eroberung.
Stuttgart 1929. (Geogr. Abh., Reihe 3, Heft 1).

BLUME, H.: Die gegenwärtigen Wandlungen in der Verbreitung von Gross- und Kleinbetrieben auf den Grossen Antillen.
Schriften d. Geogr. Inst. d. Univ. Kiel 20 (1961), p. 75–123.

BLUME, H.: Die Westindischen Inseln.
Braunschweig 1968, 2. ed. 1973.

BORCHERDT, C.: Ciudad Guayana. Kristallisationskern eines neuen Wirtschaftsraumes im südöstlichen Venezuela.
G.Z. Beihefte, Erdkundl. Wissen 41 (1975), p. 190–211.

BORCHERDT, C., K. KULINAT AND H. SCHNEIDER: Die Städte der venezolanischen Llanos.
Stuttgarter Geogr. Studien 103 (1985), p. 143–238.

BORCHERDT, C. AND S. SCHAER-GUHL: Agrarkolonisation and Agrarreform in Venezuela.
Stuttgarter Geogr. Studien 103 (1985), p. 9–141.

BORSDORF, A.: Valdivia and Osorno. Strukturelle Disparitäten und Entwicklungsprobleme in chilenischen Mittelstädten. Ein geographischer Beitrag zu Urbanisierungserscheinungen in Latainamerika.
Tübingen 1976. (Tübinger Geogr. Studien 69).

BORSDORF, A.: Grenzen und Möglichkeiten der räumlichen Entwicklung in Westpatagonien am Beispiel der Region Aisén. Natürliches Potential, Entwicklungshemmnisse und Regionalplanungsstrategien in einem lateinamerikanischen Peripherieraum.
Wiesbaden/Stuttgart 1987. (Acta Humboldtiana 11).

BRÜCHER, W.: Die Erschliessung des tropischen Regenwaldes am Ostrand der kolumbianischen Anden. Der Raum zwischen Rio Ariari und Ecuador.
Tübingen 1968. (Tübinger Geogr. Studien 28).

BRÜCHER, W.: Die moderne Entwicklung von Bogotá.
GR 21 (1969), p. 181–189.

BRÜCHER, W.: Probleme der Industrialisierung in Kolumbien unter besonderer Berücksichtigung von Bogotá und Medellin.
Tübingen 1975. (Tübinger Geogr. Studien 61).

BRÜCHER, W.: Formen und Effizienz staatlicher Agrarkolonisation in den östlichen Regenwaldgebieten der tropischen Andenländer.
G.Z. 65 (1977), p. 2–22.

BUCHHOFER, E.: Stadtplanung am Rande der Agglomeration von Mexiko-Stadt: Der Fall Nezahualcóytl.
G.Z. 70 (1982), p. 1–34.

BUCHHOFER, E.: Minatitlán (Mexiko). Zur Sozialtopographie einer Company Town in der Dritten Welt.
G.Z. 72 (1984), p. 159–178.

BÜNSTORF, J.: Die Landerschliessungs-Sukzessionen im argentinischen Gran Chaco.
Göttinger Geogr. Abh. 66 (1976), p. 91–105.

CREDNER, W.: Typen der Wirtschaftslandschaft auf den Grossen Antillen.
P.M. 89 (1943), p. 1–22.

CZAJKA, W.: Buenos Aires als Weltstadt.
In: J. H. Schultze (ed.): Zum Problem der Weltstadt.
Berlin 1959, p. 159–202. (Festschrift zum 32. Deutschen Geographentag in Berlin).

ERIKSEN, W.: Kolonisation und Tourismus in Ostpatagonien. Ein Beitrag zum Problem kulturgeographischer Entwicklungsprozesse am Rande der Ökumene.
Bonn 1970. (Bonner Geogr. Abh. 43).

EWALD, U.: Estudios sobre la hacienda colonial. Las propiedades rurales del Colegio Espirito Santo.
Wiesbaden 1976. (Das Mexiko-Projekt der DFG 9).

GANS, P.: Informelle Aktivitäten in der Altstadt Montevideos.
Verh. 45. D. G. Berlin 1985. Wiesbaden/Stuttgart 1987, p. 508–513.

GIERLOFF-EMDEN, H.-G.: Mexiko. Eine Landeskunde.
Berlin 1970.

GOLTE, W.: Das südchilenische Seengebiet. Besiedlung und wirtschaftliche Erschliessung seit dem 18. Jahrhundert.
Bonn 1973. (Bonner Geogr. Abh. 47).

GORMSEN, E.: Barquisimeto. Eine Handelsstadt in Venezuela.
Heidelberg 1963. (Heidelberger Geogr. Arb. 12).

GORMSEN, E.: Tlaxcala-Chiautempan-Apizaco. Zur Entwicklung kleiner Städte im mexikanischen Hochland.
Heidelberger Geogr. Arb. 15 (1966), p. 115–132.

GORMSEN, E.: Wochenmärkte im Bereich von Puebla. Struktur und Entwicklung eines traditionellen Austauschsystems in Mexico.
Jahrbuch für Geschichte von Staat, Wirtschaft und Gesellschaft Lateinamerikas 8 (1971), p. 366–402.

GORMSEN, E.: Die Städte im spanischen Amerika. Ein zeiträumliches Entwicklungsmodell der letzten hundert Jahre.
Erdk. 35 (1981), p. 290–303.

GORMSEN, E.: Interessentenkonflikte bei der Stadterneuerung lateinamerikanischer Kolonialstädte.
Regensburg 1986, p. 207–225. (Eichstätter Beiträge 18).

GORMSEN, E. AND K. LENZ (ed.): Lateinamerika im Brennpunkt. Aktuelle Forschungen deutscher Geographen.
Berlin 1987.

HAAS, H.-D.: Die Industrialisierungsbestrebungen auf den Westindischen Inseln unter besonderer Berücksichtigung von Jamaika und Trinidad.
Tübingen 1976. (Tübinger Geogr. Studien 68).

HEINE, K.: Mensch und geomorphodynamische Prozesse in Raum und Zeit im randtropischen Hochbecken von Puebla-Tlaxcala, Mexiko.
Verh. 41. D. G. Mainz 1977. Wiesbaden 1978, p. 390–406.

HEINE, K.: Bodenabtragung in Zentralmexiko: Messungen – Extrapolationen – Geomorphologisch-sedimentologische Befunde.
G.Z. 71 (1983), p. 28–40.

HETTNER, A.: Die Kordillere von Bogotá. Ergebnisse von Reisen und Studien.
Gotha 1892. (P.M. Erg.-Heft 104).

JÜLICH, V.: Die Agrarkolonisation im Regenwald des mittleren Rio Huallaga (Peru).
Marburg 1975. (Marburger Geogr. Schriften 63).

KLAHSEN, E.: Entwicklung und heutige Struktur der Stadt Cochabamba als Handels- und Verkehrszentrum im Ostbolivianischen Bergland.
Aachen 1983. (Aachener Geogr. Arb. 15).

KÖSTER, G.: Santa Cruz de la Sierra (Bolivien). Entwicklung, Struktur und Funktion einer tropischen Tieflandstadt.
Aachen 1978. (Aachener Geogr. Arb. 12).

KOHLHEPP, G.: Industriegeographie des nördlichen Santa Catarina (Südbrasilien). Ein Beitrag zur Geographie eines deutsch-brasilianischen Siedlungsgebietes.
Heidelberg 1968. (Heidelberger Geogr. Arb. 21).

KOHLHEPP, G.: Agrarkolonisation in Nord-Paraná. Wirtschafts- und sozialgeographische Entwicklungsprozesse einer randtropischen Pionierzone Brasiliens unter dem Einfluss des Kaffeeanbaus.
Heidelberg 1975. (Heidelberger Geogr. Arb. 41).

KOHLHEPP, G.: Strategien zur Raumerschliessung und Regionalentwicklung im Amazonasgebiet. Zur Analyse ihrer entwicklungspolitischen Auswirkungen.
Internationale Gegenwart 5 (1983), p. 175–193.

KOHLHEPP, G.: Räumliche Erschliessung und abhängige Entwicklung in Ost-Paraguay.
Lateinamerika-Studien 14 (1984), p. 203–253.

KOHLHEPP, G.: Brasilien. Beiträge zur regionalen Struktur- und Entwicklungsforschung.
Tübingen 1987. (Tübinger Geogr. Studien 93).

KROSS, E.: Mar del Plata. Ein südamerikanisches Seebad.
Göttinger Geogr. Abh. 60 (1972), p. 549–571.

KULINAT, K.: Venezuela. Industrialisierung als problemlose Strategie für kapitalreiche Entwicklungsländer?
G.R. 35 (1983), p. 333–339.

LAUER, W.: Vegetation, Landnutzung und Agrarpotential in El Salvador.
Kiel 1956. (Schriften d. Geogr. Inst. d. Univ. Kiel 16, Heft 1).

LAUER, W.: Wandlungen im Landschaftsbild des südchilenischen Seengebietes seit Ende der spanischen Kolonialzeit.
Schriften d. Geogr. Inst. d. Univ. Kiel 20 (1961), p. 227–276.

LAUER, W.: Deutsche Lateinamerikaforschung.
Jahrbuch Preussischer Kulturbesitz 1974/75, p. 71–86. [1976a].

LAUER, W.: Carl Troll – Naturforscher und Geograph.
Erdk. 30 (1976), p. 1–9. [1976b].

LAUER, W. (ed.): Landflucht und Verstädterung in Chile.
Wiesbaden 1976. (G.Z. Beihefte, Erdkundl. Wissen 42). [1976c].

LAUER, W.: Natural Potential and the Land Use System of the Kallawaya in the Upper Charazani Valley (Bolivia).
Erdwiss. Forschg. 18. Wiesbaden 1984, p. 173–196.

LICHTENBERGER, E.: The German-speaking countries.
In: R. J. Johnston and P. Claval (ed.): Geography since the Second World War. An International Survey.
Beckenham/Sydney 1984, p. 156–184.

LISS, C. C.: Die Besiedlung und Landnutzung Ostpatagoniens unter besonderer Berücksichtigung der Schafestancien.
Göttingen 1979. (Göttinger Geogr. Abh. 73).

MAASS, A.: Entwicklung und Perspektiven der wirtschaftlichen Erschliessung des tropischen Waldlandes von Peru, unter besonderer Berücksichtigung der verkehrsgeographischen Problematik.
Tübingen 1969. (Tübinger Geogr. Studien 31).

MAHNKE, L.: Zur indianischen Landwirtschaft im Siedlungsgebiet der Kallawayas (Bolivien).
Erdk. 36 (1982), p. 247–254.

MERTINS, G.: Konventionelle Agrarreformen – moderner Agrarsektor im andinen Südamerika. Die Beispiele Ecuador und Kolumbien.
In: H. Elsenhans (ed.): Agrarreform in der Dritten Welt.
Frankfurt 1979, p. 401–431.

MERTINS, G.: Typen inner- und randstädtischer Elendsviertel in Grosstädten des andinen Südamerika.
Lateinamerika-Studien 7 (1980), p. 269–296.

MIKUS, W. (ed.): Struktur- und Entwicklungsprobleme der Industrie Perus.
Heidelberg 1985. (Heidelberger Dritte Welt Studien 20).

MITTENDORFF, R.: Das Zentrum von Bogotá. Kennzeichen, Wandlungen und Verlagerungstendenzen des tertiären Sektors.
Saarbrücken/Fort Lauderdale 1984. (Sozialwiss. Studien zu internationalen Problemen 89).

MONHEIM, F.: Junge Indianer-Kolonisation in den Tiefländern Ostboliviens.
Braunschweig 1965.

MONHEIM, F.: Zwanzig Jahre Indianerkolonisation in Ostbolivien.
Wiesbaden 1977. (G.Z. Beihefte, Erdkundl. Wissen 48).

MONHEIM, F.: Die Entwicklung der peruanischen Agrarreform 1969–79 und ihre Durchführung im Departement Puno.
Wiesbaden 1981. (G.Z. Beihefte, Erdkundl. Wissen 55).

MORALES, M. AND G. SANDNER: Regiones periféricas y ciudades intermedias en Costa Rica.
San José 1982.

MÜLLER, J.: Brasilien.
Stuttgart 1984. (Klett Länderprofile).

NICKEL, H. J.: Soziale Morphologie der mexikanischen Hacienda.
Wiesbaden 1978. (Das Mexiko-Projekt der DFG 14).

NUHN, H.: Gelenkte Agrarkolonisation an der Siedlungsgrenze im tropischen Regenwald Zentralamerikas.
Göttinger Geogr. Abh. 66 (1976), p. 25–53.

NUHN, H.: Struktur und Entwicklung des Städtesystems in den Kleinstaaten Zentralamerikas und ihre Bedeutung für den regionalen Entwicklungsprozess.
Erdk. 36 (1981), p. 303–320.

NUHN, H. (ed.): Krisengebiet Mittelamerika. Zentrale Probleme, weltpolitische Konflikte.
Braunschweig 1985. (Forum Erde).

PACHNER, H.: Hüttenviertel und Hochhausquartiere als Typen neuer Siedlungszellen der venezolanischen Stadt.
Stuttgart 1982. (Stuttgarter Geogr. Studien 99).

PFEIFER, G.: Das wirtschaftsgeographische Lebenswerk Leo Waibels.
Erdk. 6 (1952), p. 1–20.

PFEIFER, G.: Brasilia.
In: A. Leidlmair (ed.): Hermann von Wissmann-Festschrift.
Tübingen 1962, p. 289–320.

PFEIFER, G.: Beiträge zur Kulturgeographie der Neuen Welt.
Berlin 1981. (Kleine Geogr. Schriften 2).

PFEIFER, G. AND G. KOHLHEPP (ed.): Leo Waibel als Forscher und Planer in Brasilien. Vier Beiträge aus der Forschungstätigkeit 1947–1950 in Übersetzung.
Wiesbaden/Stuttgart 1984. (G.Z. Beihefte, Erdkundl. Wissen 71).

POPP, K.: Privatwirtschaftliche Baulanderschliessung (fraccionamientos) im Zuge der Stadterweiterung mexikanischer Städte, aufgezeigt am Beispiel der Sadt Puebla/Pue.
Erdk. 39 (1985), p. 144–152.

RÖNICK, V.: Urbanisierung und Regionalentwicklung im Grossraum Brasilia.
G.R. 37 (1985), p. 176–182.

ROTHER, K.: Der agrarstrukturelle Wandel in Chile zwischen 1965 und 1975.
Lateinamerika-Studien 3 (1977), p. 135–153. [1977a].

ROTHER, K.: Gruppensiedlungen in Mittelchile. Erläutert am Beispiel der Provinz O'Higgins.
Düsseldorf 1977. (Düsseldorfer Geogr. Schriften 9). [1977b].

SANDER, H. J.: Sozialökonomische Klassifizierung der kleinbäuerlichen Bevölkerung im Gebiet von Puebla-Tlaxcala (Mexiko).
Bonn 1976. (Bonner Geogr. Abh. 56).

SANDNER, G.: Agrar-Kolonisation in Costa-Rica. Siedlung, Wirtschaft und Sozialgefüge an der Pioniergrenze.
Kiel 1961. (Schriften d. Geogr. Inst. d. Univ. Kiel 19, Heft 3).

SANDNER, G.: Die Hauptstädte Zentralamerikas. Wachstumsprobleme, Gestaltwandel und Sozialgefüge.
Heidelberg 1969.

SANDNER, G.: Zentralamerika und der Ferne Karibische Westen. Konjunkturen und Konflikte 1503–1984.
Stuttgart 1985.

SANDNER, G. AND H. NUHN: Das nördliche Tiefland von Costa Rica. Geographische Regionalanalyse als Grundlage für die Entwicklungsplanung.
Berlin/New York 1971. (Arbeiten aus dem Gebiet der Auslandskunde 72).

SANDNER, G. AND H.-A. STEGER: Lateinamerika.
Frankfurt 1973. (Fischer Länderkunde 7).

SCHERM, G.: Guyana und Surinam. Wirtschaftsgeographische Probleme der Rohstoffabhängigkeit bauxitexportierender Entwicklungsländer.
München 1982. (Wirtschaftswiss. Forschg. und Entwicklung 79).

SCHMIEDER, O.: Länderkunde von Südamerika, Nordamerika, Mittelamerika.
Leipzig/Wien 1932–34.

SCHMIEDER, O.: Die Neue Welt. 2 vols.
Heidelberg/München 1962/63, rev. ed. 1969.

SCHMIEDER, O. AND H. WILHELMY: Deutsche Ackerbausiedlungen im südamerikanischen Grasland. Pampa und Gran Chaco.
Leipzig 1938. (Wiss. Veröffentlichung des Museums f. Völkerkunde 6).

SCHOOP, W.: Vergleichende Untersuchungen zur Agrarkolonisation der Hochlandindianer am Andenabfall und im Tiefland Ostboliviens.
Aachen 1970. (Aachener Geogr. Arb. 4).

SCHOOP, W.: Die bolivianischen Departmentzentren im Verstädterungsprozess des Landes.
Wiesbaden 1980. (Acta Humboldtiana. Series Geogr. et Ethnogr. 7).

SCHOOP, W.: Güteraustausch und regionale Mobilität im Kallawaya-Tal (Bolivien).
Erdk. 36 (1982), p. 254–266.

SEELE, E.: Jüngere Wandlungen in der Agrarlandschaft im Hochland von Mexiko.
Verh. 37. D. G. Kiel 1969. Wiesbaden 1970, p. 536–568.

SICK, W. D.: Tropische Hochländer im Spiegel ihrer Städte. Ein Vergleich zwischen Quito (Ecuador) und Tananarive (Madagaskar).
Tübinger Geogr. Studien 34 (1970), p. 293–307.

SIEVERS, W.: Süd- und Mittelamerika.
Leipzig/Wien 1893, 3. ed. 1914.

TICHY, F. (ed.): Berichte über begonnene und geplante Arbeiten.
Wiesbaden 1968. (Das Mexiko-Projekt der DFG 1).

TICHY, F.: Ordnung und Zuordnung von Raum und Zeit im Weltbild Altamerikas. Mythos oder Wirklichkeit?
Ibero-Amerikanisches Archiv 2 (1976), p. 113–154.

TRAUTMANN, W.: Der kolonialzeitliche Wandel der Kulturlandschaft in Tlaxcala. Ein Beitrag zur historischen Landeskunde Mexikos unter besonderer Berücksichtigung wirtschafts- und sozialgeographischer Aspekte.
Essen 1983. (Essener Geogr. Arb. 5).

TROLL, C. Die Stellung der Indianer-Hochkulturen im Landschaftsaufbau der tropischen Anden.
Z.G.E. zu Berlin, 1943, p. 93–128.

TYRAKOWSKI K.: Ländliche Siedlungen im Becken von Puebla-Tlaxcala (Mexiko) und ihre Entwicklung im 19. und 20. Jahrhundert.
Berlin 1975. (Bibliotheca Ibero-Americana 21).

UMMENHOFER, S. M.: Ecuador. Industrialisierungsbestrebungen eines kleinen Agrarstaates.
Saarbrücken/Fort Lauderdale 1983. (Sozialwiss. Studien zu internationalen Problemen 78).

WAIBEL, L.: Die europäische Kolonisation Südbrasiliens.
Bonn 1955. (Coll. Geogr. 4).

WEISCHET, W.: Chile. Seine länderkundliche Individualität und Struktur.
Darmstadt 1970. (Wissenschaftliche Länderkunden 2/3).

WILHELMY H.: Südamerika im Spiegel seiner Städte.
Hamburg 1952.

WILHELMY, H.: Geographische Forschungen in Südamerika.
Berlin 1980. (Kleine Geogr. Schriften 1).

WILHELMY, H.: Welt und Umwelt der Maya. München 1981.
WILHELMY, H. AND A. BORSDORF: Die Städte Südamerikas, 2 vols. Berlin/Stuttgart 1984/85. (Urbanisierung der Erde, 3.1 and 3.2).
WILHELMY, H. AND W. ROHMEDER: Die La-Plata-Länder. Argentinien – Paraguay – Uruguay. Braunschweig 1963.

German geographical research in Africa south of the Sahara

by

Bernd Wiese (Köln)

The situation of German geographical research in Africa south of the Sahara in the mid-eighties can be described as inhomogeneous: On the one hand there are sectors of physical as well as human geography in which the results of long-established field-work are well known internationally; on the other hand some sectors are underrepresented or even non-existent. This situation can be explained by two facts: firstly, by the large gap in research activities from around 1940/41 to 1949/50 resulting from the Second World War; secondly by the fact that there are not many geographers inclined to work under severe climatic conditions and difficult administrative circumstances in tropical and southern Africa.

In addition, field-work is impossible in many African states because of civil war or unrest, the collapse of the communication and infrastructure, difficulties for research clearance or owing to restrictions on the use of maps, aerial photographs and freedom of movement which are of prime importance to geographers. One obstacle in particular hampers research in South Africa: the interpretation of a visit to South Africa as an acceptance of the apartheid regime.

Despite numerous innovative case studies, despite several widely known modern regional geographies of African countries and textbooks of general geography of Africa, despite the tremendous cartographic activities in the African Map Series Project (Afrika-Kartenwerk) of the "Deutsche Forschungsgemeinschaft" (German Research Foundation), German geographical research in Africa south of the Sahara has not yet acquired the international reputation which Eugen Wirth claims for German geographical research in the Near and Middle East (see p. 97).

German research in historical perspective

A historical perspective reveals a different picture: German *explorers* like Heinrich BARTH (from 1849 to 1855 in Northern Africa and the Sudan), Gustav NACHTIGAL (from 1863 to 1874 in the Sahara and the Sudan), Georg SCHWEINFURTH (after 1863 in the Sahara, the Sudan and Central Africa), or Herrmann von WISSMANN (crossing East and Central Africa in 1880/82) and Carl MAUCH (exploring the Zimbabwe ruins in 1871), are among the elite which emerged from the explorative phase in 19th-century Africa.

A *first phase of particular geographical research* began when in 1884 with the Conference of Berlin Germany became a colonial power acquiring territories in West, South West and East Africa. Topics of physical and economic geography were of the utmost importance and remained so even when World War I caused a suspension of fieldwork for Germans up to the 1920s. Pioneering works by Fritz JAEGER (1928b) and F. R. FALKNER (1938, 1939) dealt with the question of water resources and aridity. For Leo WAIBEL (1937), Africa was the source of raw materials for the European colonial powers. In 1928 (1928a) and 1935 respectively, Fritz JAEGER and Fritz KLUTE published well-known comprehensive regional geographies of Africa. Erich OBST (1915/1923) and Otto JESSEN (1936) carried out geomorphological field-work, the former in East Africa, the latter in Angola; they laid the foundation for extensive German geomorphological research work in Africa.

Excellent results of cartographic work during colonial times are particularly worthy of mention, e.g. the map of the semiarid interior of East Africa (north-western Tanzania) on the scale 1:300 000 by B. M. Moisel, E. Obst, and P. Sprigade (in OBST, E. 1923). Under the direction of Erich OBST, the publication of a general survey of Africa began in 1941, called "Afrika. Handbuch der praktischen Kolonialwissenschaften" (Africa. Handbook of Practical Colonial Sciences) – at a time when Lord Hailey had just published "An African Survey" in 1936.

The Second World War put a total stop to overseas research activities. It would be wrong to characterize all the geographical fieldwork in Africa before World War II as "practical colonialism". The variety and geographical distribution of topics clearly indicates that research did not concentrate on German colonies or on subjects of colonial interest.

After World War II, which caused the loss of the research capacities of universities like Breslau/Wroclaw, Leipzig or Greifswald and the publishing houses of Gotha to the West German research community, personalities like Kurt KAYSER and Joachim Heinrich SCHULTZE were among the first to resume field-work in Africa south of the Sahara. After travelling through Central and Southern Africa in 1952, Kurt Kayser visited a number of African countries from Senegal to South Africa in 1959/60. Joachim Heinrich Schultze paved the way for the continuation of German research in East Africa.

During the fifties, a new generation of geographers started fieldwork in Africa, some of them inspired by the older colleagues. Horst MENSCHING'S research work in North Africa expanded into the Sahel and the Sudan zone of West Africa. The interest in geomorphological and ecological subjects continued; the activities led to the development of a large research team at the University of Hannover.

As lecturer in geography in Ghana from 1952 to 1960, Walther MANSHARD collected the material for the first comprehensive modern geography of this West African country, a work which led to his postgraduate habilitation thesis at the University of Cologne (1959, published in 1961). Here, Kurt KAYSER was establishing a nucleus for research in Africa south of the Sahara. The habilitation thesis of Walther MANSHARD (1961) is a corner stone of the agricultural geography of tropical Africa which became one of the main subjects of German geographical research. Ernst WEIGT (1955) continued the German tradition in the field of economic and population geography of East Africa.

In the sixties, the *African Maps Series Project* (Afrika-Kartenwerk), financed by the "Deutsche Forschungsgemeinschaft" (German Research Foundation), became the main forum of German research activities in Africa. Four research teams were set up; the *North African group* under Horst MENSCHING is discussed in the article by Eugen Wirth. For *tropical West Africa,* Walther MANSHARD (Giessen, later Freiburg) led a team engaged in field-work in southern Nigeria and eastern Cameroon. *Tropical East Africa,* the region north and east of Lake Victoria, became the main research field of Joachim Heinrich SCHULTZE and his team at the Free University of Berlin. The experience of Kurt KAYSER (University of Cologne), who had been working in tropical and southern Africa since the thirties, was used to introduce a group of researchers into *southern Africa* (Mozambique, Swaziland, Transvaal).

At the end of the sixties, during the main phase of field-work, the interdisciplinary teams totalled over 40 scientists from geography and related disciplines like geology, soil sciences, botany, ethnography, African languages, history and tropical medicine. The great number of researchers, all of whom were fulltime university teachers, was one of the reasons for the delay in the publication of the results. However, the interdisciplinary approach prompted a number of young geographers of those days to become involved in long-term African research. The co-operation which began on an institutional basis in the sixties is still working on a personal basis among the "veterans".

Field-work and discussions of methodological or cartographic matters within the Afrika-Kartenwerk project led to substantial progress in geomorphology (Hanna BREMER and Heinrich ROHDENBURG 1983; Kurt KAYSER 1983/1986; Jürgen SPÖNEMANN 1980/1984). Ralph JÄTZOLD (1977b/1981) and Adolf Friedrich FABRICIUS (1984/1987) opened up new perspectives in the climatic geography of tropical and southern subtropical Africa respectively. Techniques of field-work, regional and general knowledge of agricultural geography and the geography of rural settlements in Africa as well as their cartographic representation have been expanded and refined by Klaus GRENZEBACH (1976/1984) for West Africa, by Hans HECKLAU (1976/1978) for East Africa and by Diethard CECH, Gerold RICHTER, Karl-Günther SCHNEIDER und Bernd WIESE (1982) for Southern Africa. Substantial progress has been made in population geography and mapping (Friedemann CORVINUS und Klaus GRENZEBACH 1978/1984; Werner FRICKE und Simeon H. OMINDE 1982; Diethard CECH, Gerold RICHTER, Bernd WIESE 1982/1983) as well as in the study and representation of central places (Josef MATZNETTER und Bernd WIESE 1983/1984).

The upward trend of German geographical research in Africa south of the Sahara during the sixties has been facilitated by the availability of funds, by the expansion of university staff and by the solid foundation laid by German geographers in physical and human geography. As can be seen from the research activities, the volume of knowledge and new results of innovative thinking had developed sufficiently by the midsixties for comprehensive *geographical textbooks* on Africa south of the Sahara to be envisaged again. Walther MANSHARD'S publication in the widely used "Die Grosse Illustrierte Länderkunde", Vol. II, of the Bertelsmann publishing house (1963) deserves mention, as does the even better known geography of "Africa south of the Sahara" in the series of Fischer Länderkunden, Vol. 5 (1970).

About the same time, the publishing house "Wissenschaftliche Buchgesellschaft" in Darmstadt launched a series of *regional geographical monographs* called "Wissenschaftliche Länderkunden". The volume by Horst MENSCHING on Tunisia (1968) formed a solid base for a total of 12 monographs on African countries to follow up to 1987: Manfred KUDER on Angola, 1971; Willi SCHULZE on Liberia, 1973; Manfred KUDER on Mozambique, 1975; Wolf-Dieter SICK on Madagascar, 1979; Bernd WIESE on Zaire, 1980; Ernst KLIMM, Karl-Günther SCHNEIDER and Bernd WIESE on Southern Africa, 1980; Cay LIENAU on Malawi, 1981; Jürgen SCHULTZ on Zambia, 1983; Hans Karl BARTH on Mali, 1986; Stefan von GNIELINSKI on Ghana, 1986; Bernd WIESE on the Ivory Coast, 1987.

The object of the brief overview of the history of German geographical research in Africa south of the Sahara with special emphasis on the period since World War II was to draw attention to the importance of *key personalities* of the prewar and postwar generations who succeeded in forming effective *centres of African research* at universities like Cologne, Hannover, Berlin and Hamburg. From here a new generation of researchers has emerged, to whom the tradition of African field-work is a continuing challenge; the terms of the challenge are determined by continuing social and economic developments, by innovative and expanding geographical research and, most importantly, by the ever-changing face of Africa itself.

Main subjects of German research: Physical Geography

As already indicated above one may discern a number of main topics in physical and human geography which have been studied by several generations of researchers. In *geomorphology* the formation of inselbergs and peneplains developed into one of the major controversies of the geomorphology of the tropics and the topic is still discussed even today. During the 30's, geographers like Otto JESSEN (1936) and Kurt KAYSER (In: OBST, E. and K. KAYSER, 1949) undertook considerable field-work on this question in Angola and South Africa respectively. In the 50's, the discussion on peneplanation and inselbergs emerged again when Herbert LOUIS (1964) started his work in East Africa.

When Julius BÜDEL developed his theory of the "doppelte Einebnungsfläche" (deep and differential weathering as cause of the formation of peneplains), the discussion became more controversial. Horst MENSCHING'S results from research in the Sahel and Sudan zone of West Africa demonstrated how a younger geomorphologist unbiased by established concepts could come up with new ideas. Approaching the problem through the concept of climatic geomorphology, he and others put aside the older concept of interpretation of landforms by tectonic movements or petrographic conditions.

Today two opposite positions must be distinguished: One group represented by Kurt KAYSER, Herbert LOUIS and Horst MENSCHING underlines the fact that the formation of peneplains and inselbergs takes place from the humid tropics to the semiarid outer tropics, caused by differential processes in different climatic zones. On the other hand, for Julius BÜDEL and his group represented by Hanna BREMER, the formation of this type of relief is confined to the seasonally humid tropics; in the semiarid and arid regions such reliefs are claimed to have been formed under more humid conditions in the past.

Soil erosion is another main subject of geomorphological research in Africa. In the 30's, Erich OBST had interpreted soil erosion as an indicator of the ongoing desiccation of Africa – the term desertification was not yet in use. After World War II, research work by Horst MENSCHING (1971), Hans Karl BARTH (1977) and Fouad IBRAHIM (1980, 1984), who used a broader geoecological approach and more refined methods on research analysing the desertification process, underlined the fundamental role man has been and still is playing in this context.

Since the Sahel drought from 1968 to 1973, the *study of desertification* has become an integral part of German geographical research in Africa. In this field the above-mentioned researchers Horst MENSCHING and Fouad IBRAHIM and also Hartmut LESER for southern Africa (1971, 1980) are well known internationally. They underline the fact that droughts in Africa are seasonal and quasi-natural phenomena aggravated by population increase and the extension of agricultural land; they do not agree with a one-sided politico-economic interpretation of the drought such as that given by Wulf D. SCHMIDT-WULFFEN (1985). Many proposals have been developed to combat desertification in Third World countries on the basis of the above-mentioned insight into the desertification process.

The question of *rainfall variability* and drought has been of particular interest to German geographers from the beginning of the 20th century. Substantial contributions came from Fritz JAEGER (1928b) and F. R. FALKNER (1938, 1939). With the data and methods available at the time, they examined the agroecological and climatical limits of drought-stricken areas. In the calculation and delimitation of climatically humid areas of South America and Africa and their degrees of differentiation, an interesting step forward was taken by Wilhelm LAUER (1952) who used a newly developed aridity index. In 1981, this concept was further refined for a more precise differentiation of hygrothermal climatic regions of Africa by Wilhelm LAUER and Peter FRANKENBERG. Making use of the "Potential Regional Evaporation", they developed a detailed map of the climatic classification of Africa on the scale of 1:10 million.

During his research work for the Afrika-Kartenwerk in East Africa, Ralph JÄTZOLD (1970a, 1976, 1977a) came up with an aridity index which is well suited to the evaluation of the agricultural potential of tropical areas from an agroecological point of view. This index has been largely used by the Kenyan authorities for the delineation of homogeneous agricultural planning regions.

The close connections the meteorologist Herrmann Flohn maintained with geography at university as well as at school level had a tremendous impact on *climatic geography*. They led to a more dynamic approach in interpreting the climatic zones of tropical Africa. For West Africa, Dieter KLAUS (1981) has given an example of how to follow up this method of modern climatic geography.

The *geography of soils* is clearly lagging behind the general progress in physical geography in Africa south of the Sahara. Soil science has developed into a very specific field of research, and only very few geographers are able to pursue it. The work of Arno SEMMEL (1986) who is using an integrated approach by making use of the catena principle for the interpretation of relief and soil formation has proved promising. Still an open question to soil scientists as well as to physical and agricultural geographers is the theory of Wolfgang WEISCHET, first published in 1977, that the humid tropics are disadvantaged largely by soil related factors. Although Wolfgang WEISCHET (1984) developed a more differentiated approach in the mid-eighties, his original theory is much discussed by specialists in soil science and tropical agriculture.

In sharp contrast to the situation in South America where Carl Troll had laid the foundation for a broad spectrum of *geoecological research* closely linked to vegetation geography, for Africa this approach was mainly pursued by botanists. For Southern Africa the work of Heinrich WALTER (1939, 1954) is of outstanding importance especially concerning the impact of man on vegetation. The publication of Rüdiger KNAPP (1973) is the standard handbook on the botany and vegetation geography of Africa. Knapp as a botanist was also responsible for the compilation of the map "West Africa – Vegetation" within the Afrika-Kartenwerk series (1985).

The results of the field-work performed by German geographers in Africa south of the Sahara since World War II in the field of physical geography may be summarized as follows: progress in research on desertification, geomorphology and the problem of the delineation of arid and humid regions has earned international reputation. In other sectors like hydrogeography and vegetation geography, colleagues from France and the United Kingdom lead internationally. Some suggestions to explain this situation are made in the final section.

Main subjects of German research: Human Geography

Field-work and research within human geography have concentrated on rural areas – which is not surprising due to the dominating role of agriculture in African countries. It is fascinating to observe the approach to *agricultural geography* changing from a rather formal analysis, as in traditional rural geography, to a more genetic and dynamic approach. During the 50's, Wolfgang KULS and Walther MANSHARD developed appropriate techniques. With his documentation and interpretation of planned agricultural settlements and plots, Walther MANSHARD (1961) gained international recognition. Wolfgang KULS (1958, 1963) provided geographers and cultural scientists in general with a deeper understanding of the interrelation between ethnic groups, rural economies, social structures and the agricultural landscape.

The ideas of Walther Manshard and Wolfgang Kuls were followed up by Martin BORN (1965), Ralph JÄTZOLD (1970b), Hans HECKLAU (1976/1978), Jürgen SCHULTZ (1971) and Bernd WIESE (1979). They intensified research on smallholder farming in densely populated regions of tropical Africa (see also Heinz-Dieter LUDWIG, 1967), on altitudinal belts of land-use in African mountains and highlands, as well as the development of traditional agricultural techniques, settlement patterns and land tenure under the impact of technical and social innovations.

The more recent studies on agricultural geography are characterized by a better integration of social and political factors into research. In the field of agricultural geography, German human geography has also gained international recognition on account of important regional studies (mentioned above) or significant handbooks (Walther MANSHARD 1968, 1988). The recent fieldwork on agrarian colonisation in West Africa (Wolfgang HETZEL, 1984) or on the interrelation of environment, man and land-use in African mountains and highlands (Bernd WIESE, 1985) indicate an awareness of current economic, social and ecological processes in rural Africa.

The results arrived at by German geographers within the field of *urban geography* exhibit the same high standard. The book of Walther MANSHARD (1977) on urbanization in tropical Africa became a standard work on Africa-related research in human geography; Karl-Günther SCHNEIDER und Bernd WIESE (1983) used a cultural-genetic approach for a study of urbanization in southern Africa which has been well received.

The creation and spread of *central places* as a particular problem of the developing countries was studied by Josef MATZNETTER (1966), Karl VORLAUFER (1972) and Reinhard HENKEL (1979) for East Africa, and by Wolfgang HETZEL (1974) in West Africa. While Josef Matznetter and Karl Vorlaufer traced the establishment of networks of central places from colonial times to the present day by traditional methods of field work, Reinhard Henkel made use of methods of quantitative geography for his analysis. Josef MATZNETTER and Bernd WIESE (1984) examined the historical components and strata in the growth and failure of central places in South East Africa within the Afrika-Kartenwerk, Serie South – an aspect which merits further investigation. In general, studies in urban geography in Africa south of the Sahara by German geographers have emphasized the fact that during the urbanization process in Africa, several cultural-genetic types of towns and cities developed and that there is no simple model of the African city.

Geographical research into *industrial geography* did not begin to develop until the 60's, when the countries of Africa south of the Sahara tried to follow the development of the industrialized countries. Here, Ludwig SCHÄTZL (1973) and Eike W. SCHAMP (1978) have published substantial monographs for

West and Central African countries. On the basis of several years of fieldwork, they revealed the component factors of industrialization in its positive and negative effects and discussed the industrialization process within the context of development theories.

The *communication network* plays an important role in the development process. After the pioneering study of Wolfgang HEINZE (1967), Wolfgang HETZEL (1974) and Joachim JESKE (1981) pursued questions of transportation geography in West- and Southern Africa respectively. Unfortunately, Wolfgang Hetzel's attempt at integrated research on transportation and trade in the particular instance of Togo and Benin has not been further applied – perhaps on account of the time-consuming work which had to be carried out in noisy and dusty markets, ranging from supermarkets to small indigenous shops, the informants being merchants of different nationalities who have little inclination to talk to researchers at all.

Population geography may be called a neglected part of human geographical research by Germans. While there are excellent regional studies e.g. by Jürgen BÄHR (1983) as well as impressive cartographic results within the Afrika-Kartenwerk Series (Diethard CECH et al. 1982/83; Friedemann CORVINUS and Klaus GRENZEBACH 1978/1984; Werner FRICKE and Simeon H. OMINDE 1982), there have been no further studies comparable to those published in the English or French speaking world. This is astounding as the subject is of the utmost importance to the understanding of the development process and prospects in Africa.

Some German geographers have tried to interpret *underdevelopment in Africa* as being caused by the development of the capitalistic industrial society in Europe in the same way as Walter RODNEY (1975) did in his book on African history. These authors neglect various factors inherent in African societies. It is not true that all the traditional societies have been changed, as the Africans themselves are much more aware of those values and structures which are worth preserving than we European researchers. The middle class in the independent African states is by no means simply a bourgeoisie in imitation of European models, because within 25 years of independence, new and indigenous African lifestyles and concepts have been developed.

Here, an enormous backlog of research is evident in the field of *social and political geography*. Most German researchers have to make do with all-too-brief glimpses of Africa that a two-to-three-month stay during the academic vacations affords. Long-term residence within the African context or periodic extended stays are essential for an understanding of the cultural and political scene. African colleagues can contribute to a better understanding of the African way of living and thinking and may offer their criticism of European approaches.

German Geographical Research in Africa within the National and International Context

The achievements of German geographical research in Africa south of the Sahara since World War II also have to be seen within the context of Africa-related research in general in the Federal Republic of Germany. A comparison with the geographical research based especially in France and in the United Kingdom may be instrumental in revealing the results as well as the research deficits.

Within the Federal Republic of Germany, *Hamburg* can be called a centre of research on Africa. Here, we find the "Institut für Afrika-Kunde" and the "Afrika-Verein". The History Department at the University of Hamburg has the only chair in Germany for the History of Africa south of the Sahara; the Arts Faculty also has a Department of African Languages and Cultures. In *Berlin*, departments of the Free University like the Centre for African Politology or the Department of Sociology as well as the German Institute for Development Policy are very active in African research work. Members of the Faculty of Agriculture of the Universities of *Göttingen* and *Giessen* are well known for their contributions to research in different regions of tropical and subtropical Africa, which is also carried out by the Research Unit for International Development in Agriculture at the University of Heidelberg.

In Bavaria, *Munich* was the nucleus of research on Africa; here, the IFO-Institute for Economic Research with a special section on developing countries (Dr. Axel Halbach has long been responsible for Africa) has a world-wide reputation. A Centre for African Studies has been created at the University of *Bayreuth*. Here, as well as at the Institute for Ethnology and African Languages at Munich University, research in African cultures predominates. In this field, the Institute for Ethnology and African Studies at the University of *Mainz* became well known for studies on African religious systems and structures. At the University of *Cologne*, an interdisciplinary research team is centered at the Institute for African Languages founded by Oswin Köhler, followed by Bernd Heine as director. The University of *Frankfurt/Main* is planning to expand its Africa-orientated research, making use of the long-term experience of the Frobenius Institute.

A great volume of applied research results is comprised in the reports of the *official development agencies* of West Germany, the GTZ (Gesellschaft für Technische Zusammenarbeit, Eschborn) and the KfW (Kreditanstalt für Wiederaufbau, Frankfurt). *Also private institutions* like the development agencies of the Catholic Church (Misereor, Aachen) or of the Protestant Churches (Brot für die Welt, Stuttgart) finance research activities in Africa. Unfortunately, it is rather difficult to make use of the results of these organisations as much is classified as confidential. Mention must also be made of experts outside the universities in the consulting business including geographers like Klaus Grenzebach, Giessen, and Ernst Klimm, Frechen.

The debits and credits of German Africa research in the *international perspective* can be related to preferences within the research and teaching activities at German universities as well as to the research infrastructure overseas. Here French colleagues, for instance, can make use of the international network of the ORSTOM (French Overseas Research Organisation), where housing and offices are available to researchers who sometimes stay there for years. The CEGET-research centre in Bordeaux may be called a "Max Planck Institute" for tropical geography where a team of geographers is employed on a full time basis, co-operating with partners from universities. At the same time, French colleagues have the possibility of visiting their overseas final year students and doctoral candidates periodically, as overseas visits form an integral part of the lecturing activities. Thus immediate and continuous contact with the field-work of students and overseas colleagues is possible.

The same applies to the geographers of the United Kingdom: Several university departments as well as the London-based Institute for African and Oriental Studies have continuous contact with their overseas partners. English colleagues can often be found in Africa on a long-term contract as university lecturers, a situation which is quite unusual for German geographers.

Grateful though German researchers may be for the generous financial support given to them for their field-work by the "Deutsche Forschungsgemeinschaft" (German Research Foundation), most of them are university teachers, and as such are subject to teaching obligations and travelling time limitations which are a handicap to thorough, long-term field-work in Africa.

The language factor has also to be taken into consideration: French and English colleagues can communicate with the African geographers and most of the population in their own language, whereas Germans first have to acquire a sound knowledge of the common foreign language. Even with respect to African languages, some veterans of African research of English and French origin have adapted to their guest country so well that they can communicate in a vernacular – a tremendous psychological advantage during field-work.

Here, *co-operation with African scholars* and colleagues will be of great assistance, and there are quite a few well-developed personal relations between German geographers and African colleagues from different countries. By dint of communicating with the Africans, withstanding the climatic hazards of the tropics, enjoying the ways of life of different cultures – and taking plenty of time for discussions in the field, German research in Africa south of the Sahara can be expected to produce innovative thinking and new results.

Bibliography

BADER, F.: Vegetationsgeographie – Ostafrika.
Berlin/Stuttgart 1976, 1979. (Afrika Kartenwerk, K. and B. E7).

BÄHR, J.: Die jüngere Bevölkerungsentwicklung in der Farmzone Südwestafrikas/Namibias.
Coll. Geogr. 16 (1983), p. 197–225. (Lauer-Festschrift).

BARTH, H. K.: Der Geokomplex Sahel.
Tübingen 1977. (Tübinger Geogr. Studien 71).

BARTH, H. K.: Mali.
Darmstadt 1986. (Wissenschaftliche Länderkunden 25).

BORCHERT, G.: Die Wirtschaftsräume Angolas.
Hamburg 1967. (Hamburger Beitr. zur Afrika-Kunde 6).

BORN, M.: Zentralkordofan – Bauern und Nomaden in Savannengebieten des Sudan.
Marburg 1965. (Marburger Geogr. Schriften 25).

BREMER, H. AND H. ROHDENBURG: Geomorphologie – Westafrika.
Berlin/Stuttgart 1983. (Afrika Kartenwerk, K. W2).

CECH, D., G. RICHTER AND B. WIESE: Agrargeographie – Südafrika.
Berlin/Stuttgart 1982. (Afrika Kartenwerk, K. and B. S11).

CECH, D., G. RICHTER AND B. WIESE: Bevölkerungsgeographie – Südafrika.
Berlin/Stuttgart 1982, 1983. (Afrika Kartenwerk, K. and B. S8).

CORVINUS, F.: Regionale Analyse von Volkszählungen in Südnigeria.
Giessen 1978. (Giessener Geogr. Schriften 42).

CORVINUS, F. AND K. GRENZEBACH: Bevölkerungsgeographie – Westafrika.
Berlin/Stuttgart 1978, 1984 (Afrika Kartenwerk, K. and B. W8).

FABRICIUS, A. F.: Klimageographie – Südafrika.
Berlin/Stuttgart 1984, 1987. (Afrika Kartenwerk, K. and B. S5).

FALKNER, F. R.: Die Trockengrenze des Regenfeldbaus in Afrika.
P.M. 84 (1938), p. 209–214.

FALKNER, F. R.: Beiträge zur Agrargeographie der afrikanischen Trockengebiete.
Stuttgart 1939.

FRÄNZLE, O.: Bodenkunde – Südafrika.
Berlin/Stuttgart 1983, 1984. (Afrika Kartenwerk, K. and B. S4).

FRICKE, W. AND S. H. OMINDE: Bevölkerungsgeographie – Ostafrika.
Berlin/Stuttgart 1982. (Afrika Kartenwerk, K. E8).

GNIELINSKI, ST. VON: Ghana. Tropisches Entwicklungsland an der Oberguineaküste.
Darmstadt 1986. (Wissenschaftliche Länderkunden 27).

GRENZEBACH, K.: Ländliche Siedlungen in Südostnigeria und West-Kamerun.
Diss. Giessen 1969.

GRENZEBACH, K.: Siedlungsgeographie – Westafrika.
Berlin/Stuttgart 1976, 1984. (Afrika Kartenwerk, K. and B. W9).

HECKLAU, H.: Agrargeographie – Ostafrika.
Berlin/Stuttgart 1976, 1978. (Afrika Kartenwerk, K. and B. E11).

HEINZE, G. W.: Der Verkehrssektor in der Entwicklungspolitik unter besonderer Berücksichtigung des afrikanischen Raumes.
München 1967. (IFO-Afrika-Studien 21).

HENKEL, R.: Central Places in Western Kenya. A comparative regional study using quantitative methods.
Heidelberg 1979. (Heidelberger Geogr. Arb. 54).

HETZEL, W.: Studien zur Geographie des Handels in Togo and Dahomey.
Köln 1974. (Kölner Geogr. Arb., Sonderfolge, Beitr. zur Länderkunde Afrikas 7).

HETZEL, W.: Die Erschliessung des Südwestens der Elfenbeinküste.
Afrika-Informationen. Abt. für Afrika-Forschg. am Geogr. Inst. Univ. Köln 6 (1984), p. 54–90.

IBRAHIM, F.: Desertification in Nord-Darfur.
Hamburg 1980. (Hamburger Geogr. Studien 35).

IBRAHIM, F.: Ecological imbalance in the Republic of the Sudan – with reference to desertification in Darfur.
Bayreuth 1984. (Bayreuther Geowiss. Arb. 6).

JAEGER, F.: Allgemeine Länderkunde von Afrika.
Leipzig 1928.

JAEGER, F.: Die Gewässer Afrikas.
Z.G.E. (1928), p. 158–190.

JÄTZOLD, R.: Ein Beitrag zur Klassifikation des Agrarklimas der Tropen (mit Beispielen aus Ostafrika).
Tübinger Geogr. Studien 34 (1970), p. 57–69.

JÄTZOLD, R.: Die wirtschaftsgeographische Struktur von Südtansania.
Tübingen 1970. (Tübinger Geogr. Studien 36).

JÄTZOLD, R.: Isolinien humider Monate als agrarplanerisches Mittel am Beispiel von Kenya.
Z. f. ausländische Landwirtschaft 15 (1976), p. 330–350.

JÄTZOLD, R.: Humid Month Isolines as an Aid in Agricultural Planning.
Applied Sciences and Development 9 (1977), p. 140–158.

JÄTZOLD, R.: Klimageographie – Ostafrika.
Berlin/Stuttgart 1977, 1981. (Afrika Kartenwerk, K. and B. E5).

JESKE, J.: Verkehrsgeographische Strukturwandlungen im südlichen Afrika 1975–1980.
Hamburg 1981. (Arb. aus dem Inst. für Afrika-Kunde 28).

JESSEN, O.: Reisen und Forschungen in Angola.
Berlin 1936.

KAYSER, K.: Geomorphologie – Südafrika.
Berlin/Stuttgart 1983, 1986. (Afrika Kartenwerk, K. and B. S2).

KERSBERG, H.: Vegetationsgeographie – Südafrika.
Berlin/Stuttgart 1986. (Afrika Kartenwerk, K. S7).

KLAUS, D.: Klimatologische und klimaökologische Aspekte der Dürre im Sahel.
Wiesbaden 1981. (Erdwiss. Forschg. 16).

KLIMM, E., K.-G. SCHNEIDER AND B. WIESE: Das südliche Afrika. I. Republik Südafrika–Swaziland–Lesotho.
Darmstadt 1980. (Wissenschaftliche Länderkunden 17).

KLUTE, F.: Afrika. Allgemeine Länderkunde.
Hannover 1935.

KNAPP, R.: Die Vegetation von Afrika.
Stuttgart 1973.

KNAPP, R.: Vegetationsgeographie – Westafrika.
Berlin/Stuttgart 1985. (Afrika Kartenwerk, K. W7).

KUDER, M.: Angola.
Darmstadt 1971. (Wissenschaftliche Länderkunden 6).

KUDER, M.: Moçambique.
Darmstadt 1975. (Wissenschaftliche Länderkunden 10).

KULS, W.: Beiträge zur Kulturgeographie der südäthiopischen Seenregion.
Frankfurt 1958. (Frankfurter Geogr. Hefte 34).

KULS, W.: Bevölkerung, Siedlung und Landwirtschaft im Hochland von Godjam (Nordäthiopien).
Frankfurt 1963. (Frankfurter Geogr. Hefte 39).

LAUER, W.: Humide und aride Jahreszeiten in Afrika und Südamerika und ihre Beziehung zu den Vegetationsgürteln.
Bonner Geogr. Abh. 9 (1952), p. 15–98.

LAUER, W. AND P. FRANKENBERG: Untersuchungen zur Humidität und Aridität von Afrika.
Bonn 1981. (Bonner Geogr. Abh. 66).

LESER, H.: Landschaftsökologische Grundlagenforschung in Trockengebieten. Dargestellt an Beispielen aus der Kalahari und ihren Randlandschaften.
Erdk. 25 (1971), p. 209–223.

LESER, H. (ed.): Ökologische Aspekte der Desertifikation und das Problem der Wüstenabgrenzung.
Basel 1980. (Geomethodica 5).

LIENAU, C.: Malawi.
Darmstadt 1981. (Wissenschaftliche Länderkunden 20).

LORD HAILEY: An African Survey.
London/New York/Toronto 1938.

LOUIS, H.: Über Rumpfflächen und Talbildung in den wechselfeuchten Tropen, besonders nach Studien in Tansania.
Z. f. Geomorph., N.F. 8, Sonderheft (1964), p. 43–70.

LUDWIG, H. D.: Ukara – ein Sonderfall tropischer Bodennutzung im Raum des Victoria-Sees.
München 1967. (IFO-Afrika-Studien 22).

MANSHARD, W.: Die geographischen Grundlagen der Wirtschaft Ghanas.
Köln 1971. (Kölner Geogr. Arb., Sonderfolge, Beitr. zur Länderkunde Afrikas 1).

MANSHARD, W.: Afrika – südlich der Sahara.
Gütersloh 1963, p. 233–526. (Die Grosse Illustrierte Länderkunde. Vol. 2).

MANSHARD, W.: Agrargeographie der Tropen.
Mannheim 1968. (B.I.-Hochschultaschenb. 356/356a).

MANSHARD, W.: Afrika – südlich der Sahara.
Frankfurt 1970. (Fischer Länderkunde 5).

MANSHARD, W.: Die Städte des Tropischen Afrika.
Berlin/Stuttgart 1977. (Urbanisierung der Erde 1).

MANSHARD, W.: Entwicklungsprobleme in den Agrarräumen des tropischen Afrika.
Darmstadt 1988.

MATZNETTER, J.: Die Entstehung und der Ausbau zentraler Orte und ihrer Netze an Beispielen aus Portugiesisch-Guinea und Südwestangola.
Nürnberger Wirtschafts- und Sozialgeogr. Arb. 5 (1966), p. 93–114. (Angewandte Geographie. Festschrift Erwin Scheu).

MATZNETTER, J. AND B. WIESE: Siedlungsgeographie – Südafrika.
Berlin/Stuttgart 1983, 1984. (Afrika Kartenwerk, K. and B. S9).

MATZNETTER, J. AND B. WIESE: Historische Siedlungsgeographie – Südafrika.
Berlin/Stuttgart 1984. (Afrika Kartenwerk, K. and B. S16).

MENSCHING, H.: Tunesien.
Darmstadt 1968. 3. rev. ed. 1979. (Wissenschaftliche Länderkunden 1).

MENSCHING, H.: Der Sahel in Westafrika.
Hamburger Geogr. Studien 24 (1971), p. 61–73.

OBST, E.: Das abflusslose Rumpfschollenland im nordöstlichen Deutsch-Ostafrika. Part 1 and 2. Hamburg 1915, 1923. (M.G.G. in Hamburg 29, 35).

OBST, E. (ed.): Afrika. Handbuch der praktischen Kolonialwissenschaften. 18 vols. Berlin 1941–1943.

OBST, E. AND K. KAYSER: Die Grosse Randstufe auf der Ostseite Südafrikas und ihr Vorland. Hannover 1949.

RODNEY, W.: Afrika – Geschichte einer Unterentwicklung. Berlin 1975.

SCHÄTZL, L.: Räumliche Industrialisierungsprozesse in Nigeria. Industriegeographische Analyse eines tropischen Entwicklungslandes. Giessen 1973. (Giessener Geogr. Schriften 31).

SCHÄTZL, L.: Wirtschaftsgeographie – Westafrika. Berlin/Stuttgart 1976, 1979. (Afrika Kartenwerk, K. and B. W12).

SCHAMP, E. W.: Industrialisierung in Äquatorialafrika. München 1978. (IFO-Afrika-Studien 100).

SCHMIDT-WULFFEN, W. D.: Dürre- und Hungerkatastrophen im Sahel. Kassel 1985. (Urbs et Regio 37).

SCHNEIDER, K.-G. AND B. WIESE: Die Städte des südlichen Afrika. Berlin/Stuttgart 1983 (Urbanisierung der Erde 2).

SCHULTZ, J.: Agrarlandschaftliche Veränderungen in Tansania. München 1971. (IFO-Afrika-Studien 64).

SCHULTZ, J.: Zambia. Darmstadt 1983. (Wissenschaftliche Länderkunden 23).

SCHULTZE, J. H.: Evolution und Revolution in der Landschaftsentwicklung Ostafrikas. Wiesbaden 1966. (G.Z., Beihefte, Erdkundl. Wissen 14).

SCHULZE, W.: Liberia. Länderkundliche Dominanten und regionale Strukturen. Darmstadt 1973. (Wissenschaftliche Länderkunden 7).

SEMMEL, A.: Böden des tropischen Afrika. Afrika im Spiegel neuer Forschung. Frankfurter Beitr. zur Didaktik der Geogr. 9 (1986), p. 214–222.

SICK, D.: Madagaskar. Tropisches Entwicklungsland zwischen den Kontinenten. Darmstadt 1979. (Wissenschaftliche Länderkunden 16).

SPÖNEMANN, J.: Geomorphologie – Ostafrika. Berlin/Stuttgart 1980, 1984. (Afrika Kartenwerk, K. and B. E2).

VORLAUFER, K.: Das Netz zentraler Orte in ausgewählten Räumen Tanzanias und die Bedeutung des zentralörtlichen Prinzips für die Entwicklung des Landes. Verh. 38. D.G. Erlangen-Nürnberg 1971. Wiesbaden 1972, p. 446–462.

WAIBEL, L.: Urwald, Veld, Wüste. Breslau 1921.

WAIBEL, L.: Die Rohstoffgebiete des tropischen Afrika. Leipzig 1937.

WALTER, H.: Grasland, Savanne und Busch der arideren Teile Afrikas in ihrer ökologischen Bedingtheit. Jahrb. für wiss. Botanik 87 (1939), p. 750–860.

WALTER, H.: Grundlagen der Weidewirtschaft in Südwestafrika. Stuttgart/Ludwigsburg 1954.

WALTER, H.: Klimadiagramm-Karte von Afrika. Bonn 1958.

WEIGT, E.: Europäer in Ostafrika. Köln 1955. (Kölner Geogr. Arb. 6–7).

WEISCHET, W.: Die ökologische Benachteiligung der Tropen.
Stuttgart 1977.
WEISCHET, W.: Agrarwirtschaft in den feuchten Tropen.
G.R. 36 (1984), p. 344–351.
WIESE, B.: Die Blauen Berge (Mts. Bleus, Zaire): Bevölkerung und Wirtschaft eines äquatorialafrikanischen Berglandes.
Wiesbaden 1979. (Kölner Geogr. Arb., Beitr. zur Länderkunde Afrikas 8).
WIESE, B.: Zaire. Landesnatur – Bevölkerung – Wirtschaft.
Darmstadt 1980. (Wissenschaftliche Länderkunden 15).
WIESE, B.: Seaports and Port Cities of Southern Africa.
Wiesbaden 1981.
WIESE, B.: Der Funktionswandel einer Gebirgsregion im südlichen Afrika – das Beispiel der Natalischen Drankensberge.
Z. f. Wirtschaftsgeogr. 29 (1985), p. 217–237.
WIESE, B.: Elfenbeinküste. Erfolge und Probleme eines Entwicklungslandes in den westafrikanischen Tropen.
Darmstadt 1987. (Wissenschaftliche Länderkunden 29).

German geographical research in the Middle East and North Africa

by

Eugen Wirth (Erlangen)

Introduction

In the *decades before the First World War* German Orient research* held an excellent international reputation not only amongst geographers but throughout the scientific community interested in the Near and Middle East. Of equal standing with British, French and later U.S. Americans, German scholars carried out research in the Orient with interesting questions and methods. We owe to them a surprisingly accurate and comprehensive picture of the countries and the people of the Islam world at that time and also of existing evidence of ancient-oriental, classical and Islamic cultures.

Since northern Africa and Egypt were within the political and cultural sphere of influence of France and Great Britain, the efforts of German scientists on the Orient were concentrated mainly on the territories of the *Ottoman Empire.* To name a few of the better known researchers: R. Koldewey, W. Andrae, and A. Nöldecke excavated Babylon, Assur, and Uruk; E. Herzfeld surveyed and inventorized the ancient city of Samarra. In Syria and Palestine teams of German archaeologists were involved in scientific excavations and inventories at Baalbek, Palmyra, Petra, Hatra, while in Anatolia they worked in Troja, Pergamon, Milet, Priene, Didyma, and Magnesia.

Max Freiherr von Oppenheim and Ernst Herzfeld were explorers of high international standing with wide ranging scientific interests. In Oriental research their publications are still referred to nowadays. Since 1878 many important papers were also published in the "Zeitschrift des Deutschen Palästinavereins" (Journal of the German Palestine Association). The first 40

* The German term "Orient" as used throughout this paper is a term describing the cultural region of the Islam world of the Middle East and North Africa from Morocco in the west to Afghanistan and Yemen in the east; it is not strictly synonymous with the English term Orient.

volumes in particular contain fundamental research reports on Syria and Palestine, dealing with a great variety of topics: archaeology and botany, topographic mapping of great accuracy, agricultural problems, language and settlement boundaries, geology and railway construction (c.f. A. RUPPIN 1917; G. SCHUMACHER 1886, 1897; G. DALMAN 1928–1942; M. BLANCKENHORN 1912).

German *geo-scientists* played an essential role in this research: During numerous field trips Alfred Philippson explored the geology and geomorphology of western Anatolia and Max Blanckenhorn was the first to give a comprehensive account of the geology and tectonics of Syria, the Lebanon and Palestine. Ewald BANSE (1916) wrote a first comprehensive regional geography of the Ottoman Empire and Heinrich Kiepert together with his son Richard produced maps of increasingly higher standard of accuracy of Anatolia, Syria and Mesopotamia. This was only possible through the constant close contact they maintained with the field researchers of all western countries.

Even during the First World War dignitaries of the Ottoman Empire who had received a western education co-operated with German scientists in the preservation of historic monuments: In the autumn of 1916 General Ahmed Djemal Pacha the commander in chief of the 4th Turkish Army ordered the setting up of a "bureau of inspection of monuments" for Syria, Palestine and western Arabia and attached it to the surpreme command in Damascus. The German archaeologist Theodor Wiegand was appointed Inspector General of the ancient monuments within the military sector controlled by the 4th Army. This "Deutsch-Türkisches Denkmalschutzkommando" (German-Turkish command for the preservation of monuments) inventorized the ancient and Islamic settlements and monuments in excellent monographs and maps (e.g. C. WATZINGER, K. WULZINGER 1921, 1924). For the first time aerial photography was used as a tool to systematically photograph and survey these archaeological sites. In 1918 Ahmed Djemal Pacha published a volume on "Alte Denkmäler aus Syrien, Palästina und Westarabien" (Old monuments from Syria, Palestine and western Arabia).

Between the two world wars not only the whole of northern Africa but the larger parts of the Middle East as well were in the hands of the British, French and Italians either as colonies, protectorates or as mandated territories. German researchers had little option but to concentrate on the few countries that had remained independant, namely Turkey, Persia and southern Arabia. World recession and restrictions on travel and research imposed by the Third Reich led to considerable infringements to scientific work abroad. Even worse was the emigration of many highly respected Jewish scientists.

This meant that German Orient research was able to uphold its high standard in a few research areas only. Amongst these were the geographical, ethnological and archaeological investigations of H. v. Wissmann and C. Rathjens sen. in southern Arabia and the excavations at Bogazköy (Turkey) by K. Bittel, at Persepolis (Iran) by E. Herzfeld and at Uruk (Irak) by J. Jordan and E. Heinrich. During the *Second World War* and during the following years German Orient research remained like all other research almost totally cut off from the rest of the world.

When, with the monetary reform of 1949 travelling and foreign exchange restrictions were finally lifted, geographers were amongst the first Germans who once again carried out scientific work outside Europe. There were good reasons for this: As already pointed out in the introduction to this volume, overseas research had been a focal point of interest for German geographers since Alexander von Humboldt and Ferdinand Freiherr von Richthofen, and this tradition almost made it obligatory for German geographers to work outside Europe and establish their scientific credentials and reputation with research done overseas. Travelling outside Europe was at the time still impeded by considerable financial and organizational difficulties, and almost all of those who did venture outside Europe had to do this under relatively primitive conditions. This, however, was easily overcome by the enjoyment and enthusiasm connected with the feeling of finally being "away" again, being again part of a free world. After having been subjected to sparse and mostly distorted informations on foreign countries during the Nazi regime now at last Germans wanted to see and experience the world through their own eyes.

Travel funds provided by the newly founded "Deutsche Forschungsgemeinschaft" (German Research Foundation) were rather limited during the early post war years and the Orient was therefore a "close choice" in a double sense: *North Africa and the Middle East* were relatively close to Europe spatially and therefore easy to access but they also were regions that were totally different in climate and culture and therefore of particular interest. Before the First World War these areas were looked upon as "overseas" areas; it was expensive and time consuming to travel there by ship and the trips had to be prepared carefully. Now, however, the Middle Eastern and northern African countries were within easy reach on an inexpensive budget and even by means of a normal car without four wheel drive. This situation existed virtually at the start of post war geographical overseas research and is the main reason that until today the north African and Middle Eastern countries have remained favourite research areas for German geographers.

With the exception of a few business men and medical doctors geographers were the first Germans to enter the countries of the Middle East after the war. After completing his Ph. D. Horst Mensching was able to spend several months in Morocco in 1951 and again in 1953. In 1952 Carl Rathjens sen. and Erhard Gabriel undertook a hazardous trip through the Near and Middle East in an old ramshackled Volkswagen. After a several months long struggle to gain an entry visa this author was able to spend a whole year on scientific work in Iraq in 1953. A total of DM 4,100 provided by the German Research Foundation had to cover all costs. At that time the Federal Republic of Germany did not yet have diplomatic representation in Iraq, and the German archaeologists had not yet resumed their excavations.

In 1954 Carl Rathjens jun. started on his first research trip to Afghanistan; towards the end of 1954 Wolfgang Meckelein went to the Libyan Sahara on a hazardous and for financial reasons inadequately equipped expedition sponsored by the "Gesellschaft für Erdkunde zu Berlin" (Geographical Society of Berlin). In 1956 Hans Bobek was able to travel to western Persia to resume his studies on Pleistocene glaciations, on the distribution of natural forests and woodlands and on dryland farming which he had started in 1934 and 1937. In 1957 Wolf Hütteroth personally funded a trip through eastern Anatolia, where he bought two horses and travelled through the Taurus Mountains for four months to investigate the pastoral regions and the movements associated with the transhumance of the mountain nomads. More research trips of German geographers into the Orient followed a few years later as for instance those of Adolf Leidlmair and Helmut Hahn in 1958, or Reinhard Stewig in 1959, Horst Hagedorn in 1962, Christoph Jentsch and Erwin Grötzbach in 1963, of Günter Schweizer in 1964, Ernst Giese in 1966, Eckart Ehlers in 1967, Fred Scholz in 1969, and Horst Kopp and Fouad Ibrahim in 1971.

This research tradition in the countries of the Middle East and North Africa has been carried through until today. Again and again senior students, postgraduates, Ph. D. students and young post doctorals are keen to live and work there for longer periods. This younger generation not only continues and extends the work done by the older ones but also corrects their work where necessary, puts a different emphasis and accent to it and often develops a new methodology. Therefore despite of the forced break during the Second World War and early post war years German Orient research has not lost its vigour.

Priorities of German Orient research

Up to now the older generation of German Orient Geographers has been working in the Middle East and northern Africa for three of even four decades and the second post war generation has been studying the Orient for almost two decades. The results of this work have been published in many books, numerous papers and maps and through these German Orient research has yet again achieved its international standard.

It has already been pointed out in the introduction to this volume that as a result of the historic development of our discipline there are relatively few *U.S. geographers* who have been involved in research outside their own country. Only few are prepared to undertake exploratory field work in the Orient. This is surprising in view of the fact that the older generation of *U.S. "Orientalists"* and *scholars of Islamic studies* such as S. D. Goitein, Kemal Karpat, Oleg Grabar, Bernard Lewis, Richard Ettinghausen and Charles Issawi enjoyed a high international reputation. In *Great Britain* there are famous archives and world renowned institutions of Middle East research such as the School of Oriental and African Studies of the University of London; however, the total number of scholars who are actively engaged in geographical field work in northern Africa or the Middle East is relatively small with the exception of the Department of Geography of the Centre for Middle Eastern and Islamic Studies, Durham University.

It is only in *France* where Orient research also has a broad and active basis consisting of Geographers and specialists of Islamic studies interested in present day problems, art historians and archaeologists working on the material culture, and social scientists and economists, amongst them Janine Sourdel-Thomine, Marianne Barrucand, Claude Cahen, André Raymond, Dominique Chevallier. The French Orient geographers work mainly in the Maghreb (Northwest Africa) such as the circle around Jean François Troin or in the highlands of Anatolia and Iran such as Xavier de Planhol and his disciples.

Today geographical Orient research by *German speakers* again belongs to the forefront of scientific investigations in this field; it shares this place with French Orient research focussing more on research on the material culture, and a few other nationalities that work largely on aspects of history and Islamic studies. This applies both to the number and quality of publications and to the breadth of research topics and spatial spread of research areas. This can only be touched briefly in the following account:

- In some research fields German geographers have opened up *completely new directions of research and new research priorities.* They have developed innovative themes of research and have worked with methods that had not yet been applied in the Orient. Geographical field research being related to space and dealing with the material culture as well as with the reality of every day life instead of relying solely on sources of literature and archives has led to new concepts and a better understanding of a variety of scientific problems. A few examples may demonstrate this:

  - The historic-geographical research into early distribution of settlements, settlement structures, land occupation and processes of land degradation and the resulting abandonement of land within the socio-economic context has greatly benefitted from a combination of field work and work in archives (e.g. W. HÜTTEROTH 1962, 1968, 1978; W. HÜTTEROTH and K. ABDULFATTAH 1977; M. SOYSAL 1976; E. WIRTH 1964).
  - Modern concepts have been combined with intensive field investigations on features of nomadic life style and economy (e.g. W. HÜTTEROTH 1959, 1973; J. JANZEN 1980, 1984; Chr. JENTSCH 1973; W. KRAUS ed. 1969; G. MEYER 1984; F. SCHOLZ 1974a, 1976, 1981; G. STÖBER 1978; H. von WISSMANN 1961).
  - The "Rentenkapitalismus" as an ideal model of economic behaviour and of rational action strategies in the framework of the concept of "orientalischer Wirtschaftsgeist" (the Oriental economic attitude and behaviour; e.g. H. BOBEK 1959, 1962; M. E. BONINE 1980; E. EHLERS 1978; E. WIRTH 1956, 1973).
  - The questions about the characteristic features of the Islamic city and the basic architectural and economic organization of the "bazar" as a CBD-like economic centre (e.g. K. DETTMANN 1970; E. GIESE 1980; E. GRÖTZBACH 1979; M. SCHARABI 1985; G. SCHWEIZER ed. 1977; H.-G. WAGNER 1973; E. WIRTH 1968, 1975, 1974/75).
- In other sections of Orient research where little had been known before, our *level of information and understanding* has been greatly *improved and extended* and there still is considerable room for further work. Although there are no more "blank spaces" on the maps of the Orient in a strictly cartographic sense, in terms of topics there are.
  - This applies for instance to the question of Pleistocene glaciation of the high mountains of the Near East and to the problem of the distribution of the natural vegetation, in particular the former distribution and composition of the forest that has all but disappeared (e.g. H. BOBEK 1940, 1955; H. HAGEDORN et al. 1975; W. KLAER 1962b; W. LAUER and P. FRANKENBERG 1979; H. LOUIS 1939; B. MESSERLI 1966; C. RATHJENS jun. 1972; G. SCHWEIZER 1975).
  - Great advances have also been achieved in the research on problems of agriculture and land use in Afghanistan, Iraq, Yemen, Saudi Arabia, Soviet Central Asia, Syria, Morocco, Tunisia, and Turkey (e.g. K. ABDULFATTAH 1981; H. ACHENBACH 1964, 1983; E. BETZLER 1987; H. BOBEK 1951; E. GIESE 1973; E. GRÖTZBACH 1972; Ph. HÜMMER 1984; Chr. JENTSCH 1972; H. KOPP 1981; G. MEYER 1978, 1984; H. POPP 1983; J. STADELBAUER 1983a, 1983b; H. TOEPFER 1972; E. WIRTH 1962, 1964, 1977b).

  - The close co-operation between geographers and Osmanists has led to the correction of many misunderstandings about the economic and social conditions during late Ottoman empire (e.g. H. GAUBE – E. WIRTH 1984; W.-D. HÜTTEROTH 1968, 1974, 1976; E. WIRTH 1986).
- In some sections of research *recent development processes* were analyzed for the first time. Hereby theoretical aspects of social change, modernization and the increasing dependance as well as practical consequences for government planning and development strategies were dealt with as shown by the following examples:
  - The modernization and the recent socio-economic changes of nomadic life styles and economies were investigated in Iran, Afghanistan, in Kasachstan and Baluchistan, in Syria, in Oman and in the Sheikdoms on the Arabian-Persian Gulf, in Somalia and in the Sudan (e.g. D. EHMANN 1975; E. GIESE 1983; J. JANZEN 1980; Chr. JENTSCH 1973; W. KRAUS ed. 1969; A. LEIDLMAIR 1965; G. MEYER 1982, 1984; F. SCHOLZ 1974, 1975; F. SCHOLZ ed. 1981; F. SCHOLZ and J. JANZEN ed. 1982; G. SCHWEIZER 1970; J. STADELBAUER 1984).
  - Success and failure of agrarian reforms were researched in Iran, Syria, Egypt, in Iraq, and in the Maghreb (e.g. E. EHLERS 1979a, 1986; M. HOSRY 1981; J. JANZEN 1976; U. PLANCK 1974, 1975, 1979; H. POPP 1980a, 1980b; H. SCHAMP 1972; E. WIRTH 1982).
  - The highly complex range of problems of "planning versus reality" associated with recent irrigation projects were studied in Iran, Saudi Arabia, Morocco, Egypt, Syria, Sudan, and in the Punjab (e.g. K. DETTMANN 1978; E. EHLERS 1977; G. HEINRITZ 1977; G. MEYER 1978, 1980; H.-J. PHILIPP 1976, 1982; H. POPP 1982, 1983a; E. WIRTH 1977b).
  - Worker migration and depopulation of rural areas within the framework of spatial mobility and its consequences for the areas of origin and destination have been investigated by G. HEINRITZ 1981; H. KOPP 1977a; G. MEYER 1985, 1986; G. SCHWEIZER 1980.
  - Changes in the structure of traditional urban and rural craftmanship and trade and their strategies for survival have been analyzed by A. ESCHER 1986; A. JEBENS 1983; F. SCHOLZ 1980b; E. WIRTH 1976, 1985.
  - Socio-economic structures and processes in squatter settlements, Bidonvilles and Gecekondus at the urban fringes have been studied by G. HEINRITZ 1982a; G. MEYER 1985, 1987; E. ŞEN 1972; E. WIRTH 1954.
- Over the last 30 years German geographers have published a considerable number of *regional monographies* which do not find an equivalent anywhere. Most of these monographies appeared within larger series of regional publications. These are:

- Extensive *regional geographies* ("Wissenschaftliche Länderkunden") have being been published by the publishing house "Wissenschaftliche Buchgesellschaft" in Darmstadt for Tunisia (H. MENSCHING 1968), Syria (E. WIRTH 1971), Iran (E. EHLERS 1980) and Turkey (W. HÜTTEROTH 1982). In addition to these there are publications of smaller size in a series called "Länderprofile – Geographische Strukturen, Daten, Entwicklungen" (Profiles of countries – geographical structures, data and developments). These publications focus on selected aspects mainly in the field of human geography. The countries covered are Tunisia (P. FRANKENBERG 1979), Afghanistan (D. WIEBE 1984), and the small Gulf states (F. SCHOLZ ed. 1985b). Already in 1957 H. MENSCHING published a regional geography about Morocco in the series „Geographische Handbücher" (Geographical Handbooks).
- Another series called "Ländermonographien" (Regional Monographies) consists of books by groups of co-authors usually with considerable input by geographers. So far the following books have been published: Afghanistan (W. KRAUS ed. 1972), Egypt (H. SCHAMP ed. 1977), Saudi Arabia (H. BLUME ed. 1976), and Tunisia (K. SCHLIEPHAKE ed. 1984). Less geographical input show the volumes Turkey (W. KÜNDIG-STEINER ed. 1974), Iran (H. MEHNER and U. GEHRKE ed. 1976), Kuwait (F. KOCHWASSER ed. 1961), and Pakistan (U. MALIK and A. SCHIMMEL ed. 1976). Apart from these books which constitute parts of larger series there are regional geographies on the Sultanate of Oman (F. SCHOLZ 1980), Dubai (E. GABRIEL ed. 1987), Qatar (W. RITTER 1985), and Hadramaut (A. LEIDLMAIR 1961).
- Another typical feature of German Orient geography is the great number of *monographies on individual cities and towns* of the Middle East and northern Africa with detailed maps on urban functions, architectural development and urban land use. Monographies of this kind exist about Bursa, Teheran, Isfahan, Kandahar, Kabul, Damascus, Beirut, Aleppo, Baghdad, er-Riyad, Dubai, Lahore, and about small towns in Turkey, Iran, the Lebanon and Yemen (H. AL-GENABI 1976; K. DETTMANN 1969; E. GABRIEL ed. 1987; H. GAUBE and E. WIRTH 1978; H. GAUBE and E. WIRTH 1984; H. HAHN 1964; V. HÖHFELD 1977; H. KOPP 1973; H. KOPP 1977b; J. MAYER 1979; M. MOMENI 1976; H. PAPE 1977; B. RIST 1981; L. ROTHER 1972; H. RUPPERT 1969; M. SEGER 1978; U. SPIEKER 1975; R. STEWIG 1970; D. WIEBE 1978).

The above summary of the characteristics and priorities of scientific investigations undertaken by German Orient research can only serve as a generalized guideline and does not do justice to the great variety of research topics that have been under investigation. In the following sections I shall try to focus on some particular aspects of our research activities. This focus will naturally reflect this authors personal scientific interests and involvement in Orient research. No apologies are necessary for this since personal experience and perspective can never be excluded from an account of this kind. Therefore the following sections will deal largely with research into human geography and in particular with aspects of history, socio-economics, and urban geography.

Obligations in teaching and encouraging the younger generation of researchers

The older and by now also the "middle aged" generation of German Orient geographers started their field research as individuals without the benefit of teams: Men like Carl Rathjens jun., Horst Mensching, Eugen Wirth, Wolf Hütteroth, Eckart Ehlers, Günter Schweizer and Fred Scholz had only themselves to rely on during their first expeditions. The adjustments to a new environment and completely different working conditions were difficult and not without disappointments and setbacks. It took a considerable time until one became familiar with and started to appreciate life style and culture and the behavioural patterns of the Islamic people in the Middle East and North Africa.

It is because of these experiences that the first generation of post war Orient geographers has endeavoured to introduce personally the new generation into their working areas. Joint research trips have helped to make adaptation easier for the younger generation and therefore they quickly adjusted to the very different style of living and travelling. This applies in particular to the contact with the local people which one can only gain by staying in simple hotels and other forms of accomodation, eating local food and by using local forms of transport. Outside of the larger towns one has to accept what is offered locally in any case.

The experience of the older generation with regards to scientific problems and working methods in the field as well as with respect to daily routines in dealing with people, administration and institutions has been handed on to the younger generation and they have benefitted greatly from it. This process however is not one sided and in many symposia there is an constant exchange of ideas. This rarely leads to a polarisation of opinions and most Orient geographers are personal friends. Many controversies can be resolved in mutual discussions and some may simply be due to a quick generalization.

What has been said so far shows that geographical field research depends very much on the *personal dedication of the researcher* and not like in the experimental sciences on expensive equipment and laboratories. Geographical field research is, in economic terms, work intensive but not capital intensive. For the institutions sponsoring research this has the advantage that our investigations in the Orient have never been costly.

The relatively low costs and overheads of this research, however, is also due to most of the researchers occupying tenured positions at universities or at least having medium term positions as research assistants. This means that generally our research can be conducted without or with little expenditure on salaries which often makes research projects prohibitively expensive. This also has its disadvantages: The yearly routine of teaching requires presence at the university during the term and only in exceptional cases can a stay in the field exceed 6 to 8 weeks. Even though one might be able to revisit repeatedly an area over the years this does not compare with a residence in the country stretching over a few years.

In addition German Orient geographers are usually engaged in teaching the whole spectrum of either human or physical geography. Because of a great number of students, German geographers generally have a considerable work load during exams and also spend much time supervising graduate and Ph. D. theses. There is also the work load connected with the university administration, the research foundations and the professional organizations.

Teaching a subject that is relatively popular also has its advantages for research; the great number of students gives a broad spectrum for the selection of excellent students. This enables us to encourage and stimulate qualified senior students to work in the Orient, and gives us the satisfaction that the tradition of German geographical Orient research will be continued into the future.

Interdisciplinary co-operation within the Orient research

The Islamic Orient of North Africa and the Middle East represents a relatively uniform cultural region with its own special features and a tradition that spans more than five millenia. In order to do justice to this region and its people, research into human geography has to take into account the results and the experience of neighbouring disciplines involved in Orient research. Research topics and research fields are often closely interrelated and may even overlap and it is therefore only natural that German geographers working in the Middle East and North Africa have always worked in close co-operation with colleagues from ethnology, agricultural sociology, Islamic studies, archaeology, architectural history, history of art, Ottoman studies, Assyriology, economic and social history, and research into developing countries.

Particularly close co-operation exists between geography and those neighbouring disciplines that, like geography, are involved in *research into present day problems*, undertake *empirical field research*, and concern themselves with concrete features of the *material culture*. Therefore there are close links to researchers such as Ulla Johansen and Walter Dostal (ethnology), Ulrich Planck (agricultural sociology), Heinz Gaube (Islamic studies), Klaus Kreiser (Ottoman Studies), Wolfgang Müller-Wiener, Mohamed Scharabi, Dorothée Sack, Michael Meinecke (archaeology and architectural history), Wolfgang Röllig (ancient Oriental culture) and many others. There are also numerous personal and professional contacts with representatives of the above mentioned disciplines from other western countries as well as from countries of the region itself, e.g. S. D. Goitein, Oleg Grabar, André Raymond, Halil Inalcik, Dominique Chevallier, Marianne Barrucand, Soraya Faroughi.

Because of these links the bibliographies of most geographical publications concerned with the Middle East and North Africa contain numerous references to work of neighbouring disciplines. The reverse of course also applies. This interdisciplinary exchange of ideas occasionally results in a scientific problem being recognized and even formulated in one discipline but the actual research into this problem being carried out by another discipline.

The concern of geographers for the *younger generation of researchers* is similarly not restricted to one particular discipline. The close personal and professional contact between the different Oriental disciplines allows geographical supervision and advice across disciplinary limits. This is particularly helpful to those younger colleagues from disciplines that traditionally were not engaged in empirical field research but worked in libraries, archives and museums.

The *encouragement of exploratory field research overseas* within and outside our discipline appears to be important from a research political point of view: The stock of literature and other sources in the libraries, archives and museums will be available to researchers for at least another hundred years. The objects of field research such as historical buildings and other artefacts of material culture as well as social organizations, behavioural and cultural patterns will either have disappeared entirely or at least will have been altered significantly (cf. E. WIRTH 1977a, p. LXVII-LXXI). Research into these aspects is therefore of great urgency.

Field research in the Orient as team effort

The complexity and historical depth of the cultural region of the Islamic Orient greatly exceeds the capacity of an individual researcher even if he only concerns himself with a limited problem. The co-operation within a team in the field as well as at home will bring considerable advantages. Ideal in many respects is a *team of two researchers* with different scientific interests but close personal relationship. The frictions that can never been avoided entirely between two people of different character and style of work can be reduced to a minimum; the research perspectives and prospects and methods, however, are considerably enhanced. The direct communication and discussions during the field work are highly stimulating for both parties and may result in valuable new ideas and inspirations.

Teams of this kind existed already in the decades before the First World War such as Sarre–Herzfeld, Watzinger–Wulzinger, Brünnow–Domaszewski, Krencker–Zschietzschmann, and they contributed important publications to Orient research. The same applies to the team of the two geographers C. Rathjens sen. and H. von Wissmann in the period between the two wars. Amongst the generation of present day geographers there are the teams of Fouad Ibrahim–Horst Mensching (human geography – physical geography) and Heinz Gaube–Eugen Wirth (Islamic history and architecture – urban geography).

Occasionally another kind of team work has resulted in excellent scientific work: Several post graduates being supervised by an experienced researcher investigate a similar topic in different areas or they undertake research into a range of different topics in the same area. Publications that have resulted from this kind of co-operation are the dissertations of K. DETTMANN (1969) on Damascus, by H. RUPPERT (1969) on Beirut and by H. KOPP (1973) on the towns of northern Iran all being supervised by the present author. In a similar way Horst Mensching supervised in the mid sixties three theses about Tunisia by H. ACHENBACH (1964), K. GIESSNER (1964) and G. STUCKMANN (1968) dealing with aspects of human geography, ecology and hydrogeography respectively.

The priority project "Gegenwartsbezogener Orient" (present day Orient) of the Volkswagen Foundation has enabled two other teams of post graduates under the supervision of Eckart Ehlers and Fred Scholz to undertake intensive field work in two regions of the Orient: eastern Iran and Oman. Fred Scholz also carried out a larger research project in Pakistan with the assistance of a group of graduates. Erwin Grötzbach has similarly supervised several Ph.D. theses dealing with leisure activities and tourism in several areas of the Near and Middle East.

It was also team work when, with the assistance of 22 Turkish and 15 German students and the support of the University of Istanbul, R. Stewig and D. Wiebe carried out a large poll on demographic and social aspects in the town of Bursa covering 1.350 households (R. STEWIG et al. 1980). Finally I should like to mention the work done in Morocco and North Yemen under the guidance of Klaus Müller-Hohenstein, and the research undertaken through the Berlin research station at Bardai, Chad Republic. This research is primarily concerned with problems of physical geography; since this has not been dealt with in the previous chapters it will be described here in slightly greater detail.

K. Müller-Hohenstein and his team of post graduates are engaged in *basic and applied work on vegetation.* They investigate and map the distribution of different plant species and vegetation types and try to differentiate between natural and man-made influences. In the applied work they concern themselves with questions such as the value and suitability of particular plants for the traditional uses as timber or pasture, or with problems of vegetation degradation or with the suitability of various plant communities to reduce erosion by water and wind and to improve pastoral use and dry land farming (A. AL-HUBAISHI and K. MÜLLER-HOHENSTEIN 1984; U. DEIL 1984; L. GROSSER 1985; K. MÜLLER-HOHENSTEIN 1978b; D. RAPPENHÖNER 1984; C. STRÄTZ 1985).

The Department of Geomorphology of the Free University of Berlin established in 1964/65 the *field research station Bardai* in the Tibesti Mountains, Central Sahara, Chad Republic. This station served as basis for geomorphological, geological and vulcanological as well as climatological and biogeographical work. The station was continuously occupied by scientists and postgraduates until 1968. It had to be closed down temporarily because of the political developments and the lack of security in this border area. It was reopened again from 1969 till 1971 and 1973/74 to serve as basis for research in the Central Sahara. In May 1974 it was finally closed down. Until this point of time the scientific output of the work at Bardai and the nine larger expeditions that were carried out from this base consisted of 17 doctoral and 5 habilitation theses apart from numerous publications in scientific journals (cf. D. JÄKEL et al. 1974).

Large projects and interdisciplinary working groups

Just as in the USA, in Great Britain, France and Italy there are in the Federal Republic of Germany a number of large official and semi-official institutions that concern themselves with northern Africa and the Middle East. The *"Deutsches Archäologisches Institut"* (German Archaeological Institute) maintains branches in Istanbul, Cairo, Baghdad, Teheran and Madrid (the latter beeing for Islamic art in the Maghreb) and subbranches in Damascus and Sana'a. Since 1961 the *"Deutsche Morgenländische Gesellschaft"* (German Oriental Society) owns an institute with a large library in Beirut. In Hamburg there are, in close proximity and with partly joined personnel, the *"Deutsches Orient-Institut"* (German Orient Institute), the "Nah- und Mittelostverein" (Association for the Near and Middle East), the "Dokumentations-Leitstelle Moderner Orient" (Documentation Centre Modern Orient) and the main office of the "Arbeitsgemeinschaft Vorderer Orient für gegenwartsbezogene Forschung und Dokumentation" (Working Group for research and documentation of present day Near East). In the following section, however, only those interdisciplinary projects and institutions of German Orient research will be described in which geographers have a significant input and belong to the "hard core".

The work related to the map *"Sheet Tunis"* of the *"Afrika-Kartenwerk"* (Africa mapping program) was supervised by Horst Mensching. The entire program was within the framework of a priority project of the German Research Foundation. The methodology of this mapping program has already been reviewed by B. Wiese in this volume (see p. 79). The various thematic maps of the "Sheet Tunis" (this sheet is part of the International World Map 1:1,000,000) and the accompanying explanatory notes and other publications are the result of many years of intensive field research of a team of about a dozen geographers. Up to now 13 maps have been published; they are listed in the bibliography of this review. The accompanying texts are listed under the individual authors.

Even more extensive and elaborate is the interdisciplinary project *"Tübinger Atlas des Vorderen Orients"* (Tübingen Atlas of the Middle East). According to the spokesman Wolfgang RÖLLIG (1983, p. 50 ff.) the aims of the project are as follows: "It is supposed to be a modern atlas of the entire region of the Near and Middle East between Turkey and Afghanistan, Oman and Egypt presenting reliable information of various aspects of geography as well as of the different periods of history of the Near and Middle East. The atlas is to give an overview from different angles over the entire region with the aid of generalized small scale maps and it is to pin point in detailed maps of selected areas specific problems of the region in order to stimulate further research... The entire atlas is supposed to consist of 235 map sheets at a format of 73×50 cm".

"The atlas contains two large sections: Section A (geography) with ten chapters dealing with relief and hydrology, geology, geomorphology, climate, hydrogeography, vegetation, natural regions, population and settlements, economy, and transport and communication. Section B (history) is of course structured historically. It also contains ten chapters and covers the entire region of the Near and Middle East from the stone age over the early civilizations, the classical antiquity, the middle ages to the present. By now 13 sections with a total of 141 map sheets have been published. In addition there are accompanying texts containing explanatory notes to the maps as well as scientific data which attempt to close some gaps in our knowledge". The published map sheets relevant to geography are listed in the bibliography.

This ambitious mapping project not only relies on the support of a great number of scientists from different disciplines located at the University of Tübingen; a considerable number of German Orient researchers from other universities are also involved. Amongst the geographers these are H.K. Barth (Paderborn), E. Ehlers (Bonn), E. Grötzbach (Eichstätt), W.-D. Hütteroth (Erlangen), G. Meyer (Erlangen), F. Scholz (Berlin), G. Schweizer (Köln) and E. Wirth (Erlangen). The importance of geography for this project can be assessed by the fact that the managing coordinator of the project has always been a geographer – at the beginning it was Günter Schweizer, since 1979 it has been Horst Kopp.

Horst Kopp is also leader of the interdisciplinary *Jemen-Project* which is being supported by the Volkswagen Foundation within the framework of the general project "Gegenwartsbezogene Orientforschung" (Present day related Orient Research). This project persues three goals: Firstly, the considerable shortcomings in basic research in Jemen are to be upgraded. Secondly modern development processes such as have been happening in Jemen over the last twenty years are to be monitored and analyzed. Thirdly, the project wants to make a contribution to planning for the future development of the country.

Already now we can recognize that this project has come up with interesting and highly significant results; our knowledge about southern Arabia has been considerably improved and extended. In a wide spectrum of topics and disciplines the members of the project have worked on ground water resources, natural vegetation, settlement history and vernacular, agriculture in the urban fringe, animal husbandry and timber use, weekly markets and the effects of worker migration. These results are being published in an special series called "Jemen-Studien" (Jemen Studies); five volumes have so far been published (P. BEHNSTEDT 1985; E. BETZLER 1987; H. KOPP and G. SCHWEIZER ed. 1984; G. MEYER 1986; R. STRAUB 1986).

The "*Arbeitsgemeinschaft Afghanistan*" (Working Group Afghanistan) was founded in 1966 in Heidelberg. This working group encompasses researchers from different disciplines and professionals involved in development aid projects. All worked in Afghanistan for a considerable time and therefore they have close links with this country. Starting point for this working group was the partnership between the Universities of Bonn, Cologne and Bochum with the University of Kabul. The working group aims at maintaining a close contact with Afghan scientists and to organize joint meetings and symposia, and to conduct interdisciplinary research projects. It also wants to inform the German speaking public about Afghanistan. The core of this rather loose working group consists of geographers, ethnologists and botanists all being scientists involved in field research (cf. C. RATHJENS jun. 1984).

A few years after its foundation the working group was able to publish a two volume "Bibliographie der Afghanistan-Literatur 1945–1967" (Bibliography of the Afghanistan Literature 1945–1967; Hamburg 1968/69). The book series "Afghanische Studien" (Afghan Studies; 18 volumes between 1969 and 1979) and the book "Afghanistan. Natur, Geschichte und Kultur, Staat, Gesellschaft und Wirtschaft" (Afghanistan. Nature, History and Culture, State, Society and Economy; W. KRAUS ed. 1972) which is in its third edition can be attributed to the initiative of the working group. The group also published between 1974 (volume 1) and 1982 (volume 8) a scientific journal called "Afghanistan-Journal".

Every two to three years the working group meets in symposia; the papers are usually published. Since 1974 the chairman of the working group has been a geographer; from 1974 to 1982 it was Carl Rathjens jun. (Saarbrücken), from 1982 to 1986 Christoph Jentsch (Mannheim) and since 1986 Erwin Grötzbach (Eichstätt). After the military coup in 1978 and the invasion of Sovjet troops in 1979 western scientists can no longer conduct field work in Afghanistan. The working group therefore has concentrated since on working up and publishing the information and material collected before 1978.

In 1966 some ethnologists interested in the Near and Middle East met in St. Augustin near Cologne. They decided to meet regularly in working sessions together with human geographers also engaged in work in the Orient and to discuss common scientific problems. In 1972 this led to the foundation in Heidelberg of the *"Völkerkundlich-kulturgeographischer Arbeitskreis gegenwartsbezogener Orient"* (Working Group Ethnology and Human Geography for Present Day Orient). Of the foundation members the following persons are still working actively in the group: the ethnologists Ulla Johansen, Ursula Reinhard and Walter Dostal, and the human geographers Wolf Hütteroth and Eugen Wirth. Over the years some 35 ethnologists and geographers all engaged in field work in the Middle East and North Africa joined the working group. The members range from young post doctorals to old professors near retiring age.

While the working group Afghanistan encompasses scientists from different disciplines and professionals engaged in practical work the "Working Group Present Day Orient" only consists of academics from closely related disciplines namely ethnology, human geography, agricultural sociology and present day Islamic studies. While one of the main aims of the "Working Group Afghanistan" lies in public relation work and development advice, the "Working Group Present Day Orient" is more scientifically oriented and the discussions are mostly concerned with scientific problems; there is no attempt to present the group to the public.

The constituent meeting of the working group in 1972 in Heidelberg served as a first exchange of information about the level of ethnological and cultural geographical research and the possibilities and limitations of field work in various countries of the Orient. The following meetings then discussed more specialized topics. These are:

1973 in Erlangen: Houses and settlements in the Near and Middle East.
1974 in Bern: The Oriental city.
1975 in Tübingen: The market in the Near and Middle East.
1976 in St. Augustin: Problems of the integration of marginal groups in the Near and Middle East.
1977 in Hohenheim: Agricultural development in the Near and Middle East.
1978 in Marburg: Social groups in the Near and Middle East.
1981 in Vienna: Recent changes of traditional forms of economic and social organization in the Near and Middle East.
1984 in Cologne: The recording of social reality in the Near East; theoretical models and empirical research strategies.
1987 in Bayreuth: The changes of the agricultural community in the Near and Middle East.

During these meetings many new ideas, concepts and research results were presented for the first time and discussed critically in this small circle. They were later published in books or journals and they have usually found general acceptance. The younger generation of researchers also has an opportunity in this group to present their research findings to an expert audience and also to establish personal links with other researchers. Without a fixed organizational framework and without publicity this working group has managed to become a community of scientists of high standard.

The role of German Orient geography in the framework of an English speaking scientific community

The fact that *publications written in German* do not enjoy a wide distribution outside of Central Europe does not amount to a serious handicap for the position of German geographical Orient research internationally; however, it does constitute more than just a minor blemish. In the natural sciences therefore German scientists usually publish in English even in German journals. This concession to the Anglo Saxon language community appears to be acceptable and sensible if English speaking scientists are at the research front. The language of the exact sciences also is rather simple and uncomplicated and one does not have to be a linguistic genius to publish in a language other than the mother tongue.

The question however that comes to mind: Is it justified to use English if English speaking researchers are *not* generally at the forefront of research? Even people who speak English well may have difficulties to express their thoughts and intentions as precisely and as differentiated as in their own tongue. This applies in particular to human geographical research which concerns itself with human actions, thoughts and decisions. These should be presented in a language that the author fully commands and where he can express himself in all the detail and subtleness he feels necessary. Human geography is a cultural and social science and therefore its results are best presented in the language that is the researchers mother tongue. Translations always lead to a loss of information and this can be considerable (c.f. Bruno S. FREY, "Die Zeit", 21. 6. 1985). Therefore it has become the normal procedure during meetings of Orient researchers from Continental Europe to present their research findings in their native language and usually they are well understood by their foreign colleagues. The German Orient geographers have reason to be self confident and can look at their scientific work with pride. Most of them understand English and French quite well; but why should they not present their results in German at international meetings?

The cooperation of German Orient research with *French scientists* only very rarely is restricted by language barriers. The majority of Orient geographers on either side are able to read and understand easily the language of their neighbours, many also speak it fluently. Therefore it is quite natural that during geographical Orient symposia in the Federal Republic, French colleagues are invited and present papers. The same applies to German geographers attending meetings in France. The presently best known French Orient geographer Xavier de Planhol works in close scientific and personal contact with German colleagues. The most prominent French geographer, the nestor of French Orient geography Jean Dresch who was president of the IGU from 1972 to 1976 speaks all the common European languages.

French geographers – Jean Dresch and Xavier de Planhol, Jean-François Troin and Marcel Bazin, Daniel Balland, Bernard Hourcade and André Burgey – have worked for the last 30 years in the highlands of the Near and Middle East and in the Maghreb as well as in Syria and Lebanon and they have come up with excellent research achievements in the field of human geography. Being able to understand and speak three languages, French, German and English and often Spanish and/or Italian as well, the *Orient geographers of Continental Europe* have a *considerable advantage* over their British and

American colleagues who usually only understand, use and quote English language publications.

Therefore publications about the Orient of English speaking geographers, ethnologists and social scientists are sometimes well behind the main stream of research and even appear occasionally provincial. The authors of these publications simply can not put their research results into the context and perspectives with similar work done by French or German Orient researchers. How little the English speaking scientific community knows about the Orient research conducted by Continental Europeans is shown by two well known recent publications of W. B. FISHER (1981, 1983). Fisher reports rather well intentioned about the prominent position internationally of German geographical Orient research. Since, however, he obviously lacks adequate knowledge about German he has no grasp of the content of the publications and his selection of references is rather haphazard to say the least.

In sharp contrast to these younger "specialists" of Anglo Saxon Orient research are the "great old men" of Oriental and Islamic studies, Near Eastern archaeology and art history of Great Britain and North America, men like Bernard Lewis, S. D. Goitein, Oleg Grabar, Kemal Karpat, Richard Ettinghausen and Gustave von Grunebaum. They all are scientists of wide interests and all speak or spoke foreign languages, at least French and German. Those that are still alive maintain close contacts to the older generation of German Orient geographers. For the younger Anglo Saxon colleagues the German Orient research is at best only known through the English translations in "Applied Geography and Development" and through reports in the "Middle East Studies Association" (MESA) (c.f. E. EHLERS 1985), often however only through general review papers and literature reviews in journals that are available in their countries.

For the *Arabian speaking communities* a large volume containing 32 papers by German scientists dealing with a variety of geographical topics about North Africa and the Middle East was published for the International Geographical Congress held in 1984 in Paris (E. WIRTH ed. 1983). All the papers were translated into Arabic which was made possible through the support of the Volkswagen Foundation. This publication in Arabic was intended as a small token of gratitude and an expression of the close bond of friendship that German geographers have developed with their colleagues in the Arab world. As paradox as it may seem: It was through this publication that some younger "Orientalists" and Arabists from Great Britain and the U.S.A. who can read Arabic learned about the research of German geographers in the Middle East.

Future prospects

Having travelled and carried out research in the Middle East and North Africa for decades and having benefitted from the generous hospitality of the people there one may easily get a bad conscience. Without the active, generous help and co-operation of the local population, colleagues, administration and business it would be impossible for the forcign scientist to work in the Orient. Essential always has been the indirect tacit agreement of the "man in the street" who may be the actual object of research. However, it would be a serious abuse against humanity if the land and above all the people of the Orient were only treated as "objects", as "material" or simply as "means to an end" as supplier of data for scientific publications to the personal gain of the researcher. Such a "scientific imperialism" fortunately is foreign to most German Orient geographers; over the years many have developed *close personal ties* with the people and the countries they have worked in and many feel very much at home in the area where they have done intensive field work.

As a visible sign of these bonds with the people of the Middle East and North Africa and as small "repayment" for help that we received in the Orient we should endeavour to involve even more than before local scientists and above all the younger generation of geographers in our research and field work. Such a *co-operation* will assist our colleagues in establishing and strengthening scientific research and in particular field research. One example of this kind of co-operation was the significant impact by Germans on the development of science in Turkey where German emigrants tought at Turkish universities during the Nazi era. Amongst them such well known personalities as the economist Fritz Baade, the art historian Kurt Erdmann, the geographer Herbert Louis, the zoologist Curt Koswig and last but not least Ernst Reuter who after the war became Lord Major of Berlin.

We are also able to influence and promote Orient research by supporting and supervising Ph.D. students and young scientists from the Middle-Eastern countries at our universities. This will help to further improve the standard and prestige of Orient research in their own countries. Up to now there are only about a dozen geographers from Middle East and North Africa who have gained their Ph. D. in Germany and who occupy university positions as lecturers or professors in their home countries. Two of the most active German university teachers in this respect are the geographer *Wolf Hütteroth* (Erlangen) and the agricultural sociologist *Ulrich Planck* (Stuttgart-Hohenheim). Without doubt, however, more could be done and it is to be hoped that with support from either side and with the assistance of scholarships and other funding programs the co-operation between us and the Oriental countries will further improve to our mutual benefit.

Bibliography

A. German fieldwork before Second World War.
(Publications of authors who had done fieldwork before 1939)

BANSE, E.: Die Türkei. Eine moderne Geographie.
Braunschweig, 2. ed. 1916.

BLANCKENHORN, M.: Grundzüge der Geologie und physikalischen Geographie von Nord-Syrien. Eine geologisch-geographische Skizze.
Berlin 1891.

BLANCKENHORN, M.: Kurzer Abriss der Geologie Palästinas. Begleitworte zu einer neuen geologischen Karte von Palästina.
Z. des Dt. Palästinavereins 35 (1912), p. 113–139.

BOBEK, H.: Forschungen im Zentralkurdischen Hochgebirge zwischen Van- und Urmia-See (Südostanatolien und Westazerbaican).
P.M. 84 (1938), p. 152–162, 215–228.

BOBEK, H.: Die gegenwärtige und eiszeitliche Vergletscherung im zentralkurdischen Hochgebirge (Osttaurus, Ostanatolien).
Z. f. Gletscherkunde 27 (1940), p. 50–87.

BOBEK, H.: Soziale Raumbildungen am Beispiel des Vorderen Orients. Verh. 27. D.G. München 1948. Landshut 1950/51, p. 193–206.
Arabic translation in: E. Wirth (ed.): Deutsche Geographische Forschung im Orient. Erlangen/Beirut 1983, p. 11–28.

BOBEK, H.: Die Verbreitung des Regenfeldbaus in Iran.
Geogr. Studien. Wien 1951, p. 9–30. (Festschrift für Johann Sölch).

BOBEK, H.: Beiträge zur klima-ökologischen Gliederung Irans.
Erdk. 6 (1952), p. 65–84.

BOBEK, H.: Klima und Landschaft Irans in vor- und frühgeschichtlicher Zeit.
Geogr. Jahresber. aus Österreich 25 (1955), p. 1–42.

BOBEK, H.: Die Hauptstufen der Gesellschafts- und Wirtschaftsentfaltung in geographischer Sicht.
Die Erde 90 (1959), p. 259–298.

BOBEK, H.: Iran. Probleme eines unterentwickelten Landes alter Kultur.
Frankfurt/Berlin/Bonn 1962. (Themen zur Geogr. und Gemeinschaftskunde).

BOESCH, H.: Erdöl im Mittleren Osten.
Erdk. 3 (1949), p. 68–82.
Arabic translation in: E. Wirth (ed.): Deutsche Geographische Forschung im Orient. Erlangen/Beirut 1983, p. 29–48.

DALMAN, G.: Arbeit und Sitte in Palästina. 7 in 8 vols.
Gütersloh 1928–1942. Reprint Hildesheim 1964. (Schriften des Dt. Palästina-Instituts, vols. 3, 5–10).

HEINRICH, E.: Moderne arabische Gehöfte am unteren Euphrat und ihre Beziehungen zum "Babylonischen Hofhaus".
Mitt. der Dt. Orient-Ges. 82 (1950), p. 19–46.

HEINRICH, E.: Haus. B Archäologisch.
In: Reallexikon der Assyriologie und Vorderasiatischen Archäologie.
Vol. 4, Berlin 1973, p. 176–220.

HERZFELD, E.: Archäologische Mitteilungen aus Iran.
Berlin 1 (1929/30) – 9 (1938).

HERZFELD, E.: Damascus. Studies in Architecture I–IV.
Ars Islamica 9 (1942), p. 1–53; 10 (1943), p. 13–70; 11/12 (1946), p. 1–71; 13/14 (1948), p. 118–138.
LOUIS, H.: Das natürliche Pflanzenkleid Anatoliens, geographisch gesehen.
Stuttgart 1939. (Geogr. Abh. 3, Heft 12).
LOUIS, H.: Probleme der Kulturlandschaftsentwicklung in Inneranatolien.
Erdk. 2 (1948), p. 146–151.
LOUIS, H.: Die Bevölkerungsverteilung in der Türkei 1965 und ihre Entwicklung seit 1935.
Erdk. 26 (1972), p. 161–177.
OPPENHEIM, M. VON: Vom Mittelmeer zum Persischen Golf.
2 vols. Berlin 1899, 1900.
OPPENHEIM, M. VON: Die Beduinen.
2 vols. Leipzig 1939, 1943.
PHILIPPSON, A.: Das westliche Klein-Asien auf Grund eigener Reisen.
Z.G.E. zu Berlin (1904), p. 257–273.
PHILIPPSON, A.: Reisen und Forschungen im westlichen Kleinasien. Part 1–4.
P.M., Erg.-Heft 167 (1910), 172 (1911), 177 (1913), 180 (1914), 183 (1915).
RATHJENS, SEN., C.: Löss in Tripolitanien.
Z.G.E. zu Berlin (1928), p. 211–228.
RATHJENS SEN., C.: Die Pilgerfahrt nach Mekka – Von der Weihrauchstrasse zur Ölwirtschaft.
Hamburg 1948. (Hamburgische Abh. zur Weltwirtschaft).
RATHJENS SEN., C. AND H. VON WISSMANN: Sanaa. Eine südarabische Stadtlandschaft.
Z.G.E. zu Berlin (1929), p. 329–353.
Arabic translation in: E. Wirth (ed.): Deutsche Geographische Forschung im Orient. Erlangen/Beirut 1983, p. 49–69.
RATHJENS SEN., C. AND H. VON WISSMANN: Rathjens-von Wissmannsche Südarabien-Reise. Vol. 3: Landeskundliche Ergebnisse.
Hamburg 1934. (Abh. aus dem Gebiet der Auslandskunde 40).
RUPPIN, A.: Syrien als Wirtschaftsgebiet.
Berlin 1917.
SARRE, F. AND E. HERZFELD: Archäologische Reise im Euphrat- und Tigris-Gebiet.
4 vols. Berlin 1911, 1920.
SCHUMACHER, G.: Der Dscholan. Zum ersten Male aufgenommen und beschrieben.
Z. des Dt. Palästinavereins 9 (1886), p. 165–368.
SCHUMACHER, G.: Das südliche Basan. Zum ersten Male aufgenommen und beschrieben.
Z. des Dt. Palästinavereins 20 (1897), p. 65–227.
WATZINGER, C. AND K. WULZINGER: Damaskus. Die antike Stadt.
Berlin/Leipzig 1921. (Wiss. Veröffentlichungen des deutsch-türkischen Denkmalschutzkommandos 4).
WENZEL, H.: Forschungen in Inneranatolien. II. Die Steppe als Lebensraum.
Kiel 1937. (Schriften des Geogr. Inst. der Univ. Kiel VII, 3).
WISSMANN, H. VON: Arabien.
In: F. Klute (ed.): Vorder- und Südasien in Natur, Kultur und Wirtschaft.
Potsdam 1937, p. 178–211. (Handbuch der Geogr. Wiss.).
WISSMANN, H. VON: Stellung und Bedeutungswandel des Orients in den Lebensräumen der Alten Welt.
Z.G.E. zu Berlin (1942), p. 353–368.
WISSMANN, H. VON: Bauer, Nomade und Stadt im islamischen Orient.
In: R. Paret (ed.): Die Welt des Islam und die Gegenwart.
Stuttgart 1961, p. 22–63.

WULZINGER, K. AND C. WATZINGER: Damaskus. Die islamische Stadt.
Berlin/Leipzig 1924. (Wiss. Veröffentlichungen des deutsch-türkischen Denkmalschutz-kommandos 5).

B. German fieldwork after Second World War

(Publications of authors who started fieldwork after 1950)

ABDULFATTAH, K.: Mountain Farmer and Fellah in 'Asīr, Southwest Saudi Arabia. The Conditions of Agriculture in a Traditional Society.
Erlangen 1981. (Erlanger Geogr. Arb., Sonderband 12).

ABDULFATTAH, K.: Stammesstruktur, Siedlung und Agrarlandschaft im Hochland von Asir.
In: E. Wirth (ed.): Deutsche Geographische Forschung im Orient. Erlangen/Beirut 1983, p. 164–193. (In Arabic).

ACHENBACH, H.: Die Halbinsel Cap Bon. Strukturanalyse einer mediterranen Kulturlandschaft in Tunesien.
Hannover 1964. (Jahrb. G.G. Hannover).

ACHENBACH, H.: Agrargeographische Entwicklungsprobleme Tunesiens und Ostalgeriens. Exemplarische Strukturanalyse ausgewählter Reform- und Traditionsräume zwischen Mittelmeerküste und Nordsahara.
Hannover 1971. (Jahrb. G.G. Hannover).

ACHENBACH, H.: Agricultural Geography.
Berlin/Stuttgart 1983. (Afrika Kartenwerk, B. N 11).

AL-GENABI, HASHIM K. N.: Der Suq (Bazar) von Bagdad. Eine wirtschafts- und sozialgeographische Untersuchung.
Erlangen 1976. (Erlanger Geogr. Arb. 36).

AL-HUBAISHI, A. AND K. MÜLLER-HOHENSTEIN: An introduction to the vegetation of Yemen. Ecological basis, floristic composition, human influence.
Eschborn 1984.

ARNOLD, A.: Economic Geography.
Berlin/Stuttgart 1980. (Afrika Kartenwerk, B. N 12).

ARNOLD, A.: Transportation Geography.
Berlin/Stuttgart 1981. (Afrika Kartenwerk, B. N 13).

BARTH, H. K.: Probleme der Wasserversorgung in Saudi-Arabien.
Wiesbaden 1976. (G.Z., Beihefte, Erdkundl. Wissen 45).

BECKER, H. AND H. KOPP (ed.): Resultate aktueller Jemen-Forschung. Eine Zwischenbilanz.
Bamberg 1978. (Bamberger Geogr. Schriften 1).

BECKER, H., V. HÖHFELD AND H. KOPP: Kaffee aus Arabien. Der Bedeutungswandel eines Weltwirtschaftsgutes und seine siedlungsgeographische Konsequenz an der Trockengrenze der Ökumene.
Wiesbaden 1979. (G.Z., Beihefte, Erdkundl. Wissen 46).

BEHNSTEDT, P.: Die nordjemenitischen Dialekte. Teil 1: Atlas.
Wiesbaden 1985. (Jemen-Studien 3).

BETZLER, E.: Sozialer Umbruch und Kulturlandschaftswandel in Südarabien. Agrargeographische Untersuchungen im Umland von Ṣan'ā' (Arabische Republik Jemen).
Wiesbaden 1987. (Jemen-Studien 5).

BLUME, H. (ed.): Saudi-Arabien. Natur, Geschichte, Mensch und Wirtschaft.
Tübingen/Basel 1976. (Ländermonographien 7).

BONINE, M. E.: Yazd and its hinterland.
Marburg 1980. (Marburger Geogr. Schriften 83).

BORN, M.: Zentralkordofan. Bauern und Nomaden in Savannengebieten des Sudan.
Marburg 1965. (Marburger Geogr. Schriften 25).

BORN, M.: Anbauformen an der agronomischen Trockengrenze Nordostafrikas.
G.Z. 55 (1967), p. 243–278.

BRECHTEL, R. AND W. ROHMER: Pedology.
Berlin/Stuttgart 1980. (Afrika Kartenwerk, B. N 4).

CZAPEK, F.-M.: Binnenerholungsverkehr im türkischen Schwarzmeergebiet.
Regensburg 1983. (Eichstätter Beitr., Abt. Geogr. 6).

DEIL, U.: Zur Vegetation im Zentralen Rif (Nordmarokko) – unter besonderer Berücksichtigung der Zedernwälder und ihrer Ersatzgesellschaften.
Vaduz 1984. (Dissertationes Botanicae 74).

DEIL, U. AND K. MÜLLER-HOHENSTEIN: Beiträge zur Vegetation des Jemen. I. Pflanzengesellschaften und Ökotopgefüge der Gebirgstihamah am Beispiel des Beckens von At Tur (J.A.R.).
Phytocoenologia 13 (1985), p. 1–102.

DETTMANN, K.: Damaskus. Eine orientalische Stadt zwischen Tradition und Moderne.
Erlangen 1969. (Erlanger Geogr. Arb. 26).

DETTMANN, K.: Zur Variationsbreite der Stadt in der islamisch-orientalischen Welt. Die Verhältnisse in der Levante sowie im Nordwesten des indischen Subkontinents.
G.Z. 58 (1970), p. 95–123.
Arabic translation in: E. Wirth (ed.): Deutsche Geographische Forschung im Orient. Erlangen/Beirut 1983, p. 426–443.

DETTMANN, K.: Die britische Agrarkolonisation im Norden des Industieflandes. Der Ausbau der Kanalkolonien im Fünfstromland.
M. Fränkische G.G. 23/24 (1976/77). Erlangen 1978, p. 375–411.

EHLERS, E.: Südkaspisches Tiefland (Nordiran) und Kaspisches Meer. Beiträge zu ihrer Entwicklungsgeschichte im Jung- und Postpleistozän.
Tübingen 1971. (Tübinger Geogr. Studien 44).

EHLERS, E. (ed.): Beiträge zur Physischen Geographie Irans.
Marburg 1974. (Marburger Geogr. Schriften 62).

EHLERS, E.: Traditionelle und moderne Formen der Landwirtschaft in Iran.
Marburg 1975. (Marburger Geogr. Schriften 64). [1975a].

EHLERS, E.: Die Stadt Bam und ihr Oasen-Umland/Zentraliran.
Erdk. 29 (1975), p. 38–52. [1975b].

EHLERS, E.: Kulturgeographische und agrarsoziale Aspekte grosser Bewässerungsprojekte in Trockengebieten: Das Dez-Irrigation-Projekt/Iran.
Z. f. Bewässerungswirtschaft 12 (1977), p. 127–149.
Arabic translation in: E. Wirth (ed.): Deutsche Geographische Forschung im Orient. Erlangen/Beirut 1983, p. 248–262.

EHLERS, E.: Rentenkapitalismus und Stadtentwicklung im islamischen Orient. Beispiel Iran.
Erdk. 32 (1978), p. 124–142.

EHLERS, E.: Die iranische Agrarreform – Voraussetzungen, Ziele und Ergebnisse.
In: H. Elsenhans (ed.): Agrarreform in der Dritten Welt.
Frankfurt/New York 1979, p. 433–470. [1979a].

EHLERS, E. (ed.): Beiträge zur Kulturgeographie des islamischen Orients.
Marburg 1979. (Marburger Geogr. Schriften 78). [1979b].

EHLERS, E. Iran. Grundzüge einer geographischen Landeskunde.
Darmstadt 1980. (Wissenschaftliche Länderkunden 18).
(Iranian translation by Sahab Geographic and Drafting Institute, 1365).

EHLERS, E.: Teppichmanufaktur und Teppichhandel in Arak (Farahan), Iran.
Islam 59 (1982), p. 221–253.

EHLERS, E.: Zur baulichen Entwicklung und Differenzierung der marokkanischen Stadt: Rabat – Marrakech – Meknes.
Die Erde 115 (1984), p. 183–208.

EHLERS, E.: German geography of the Middle East: Trends and prospects.
Middle East Studies Association Bulletin vol. 19, no. 2 (1985), p. 183–195.

EHLERS, E.: Agrarreform und agrarsozialer Wandel im islamischen Orient. Entwicklungsprobleme ländlicher Räume zwischen Modernisierung und Reislamisierung.
In: W. Andres, E. Buchhofer and G. Mertins (ed.): Geographische Forschung in Marburg. Marburger Geogr. Schriften 100 (1986), p. 19–38.

EHMANN, D.: Bahtiyāren – Persische Bergnomaden im Wandel der Zeit.
Wiesbaden 1975. (Beihefte zum Tübinger Atlas des Vorderen Orients, Reihe B, no. 15).

EL-SHAGI, EL-SHAGI: Neuordnung der Bodennutzung in Ägypten.
München 1969. (IFO Afrika-Studien 36).

ESCHER, A.: Grundeigentumsstruktur und Grundstücksmobilität in Taza/Marokko. Der Prozess der Entkolonisierung im Spiegel des innerstädtischen Grundeigentums.
M. Fränkische G.G. 27/28 (1980/81). Erlangen 1982, p. 455–581.

ESCHER, A.: Studien zum traditionellen Handwerk der orientalischen Stadt. Wirtschafts- und sozialgeographische Strukturen und Prozesse anhand von Fallstudien in Marokko.
Erlangen 1986. (Erlanger Geogr. Arb. 46).

FISCHER, W.: Periodische Märkte im Vorderen Orient, dargestellt an Beispielen aus Nordostanatolien (Türkei) und Nordafghanistan.
Hamburg 1984. (Mitt. des Dt. Orient-Inst. 24).

FISHER, W. B.: Progress in the Middle East.
Progress in Human Geography. An international review of geographical work in the social sciences and humanities. 5 (1981), p. 432–438; 7 (1983), p. 429–435.

FRANKENBERG, P.: Florengeographische Untersuchungen im Raume der Sahara. Ein Beitrag zur pflanzengeographischen Differenzierung des nordafrikanischen Trockenraumes.
Bonn 1978. (Bonner Geogr. Abh. 58).

FRANKENBERG, P.: Tunesien. Ein Entwicklungsland im maghrebinischen Orient.
Stuttgart 1979. (Klett Länderprofile).

FRANKENBERG, P.: Zur Landschaftsdegradation in Südosttunesien.
Wiesbaden 1983. (Akad. d. Wiss. und d. Literatur. Mainz. Abh. d. Math.-Naturwiss. Kl., no. 4).

GABRIEL, E.: Zur Erdölwirtschaft am Persischen Golf. Eine wirtschaftsgeographische Betrachtung.
M. Fränkischen G.G. 13/14 (1966/67). Erlangen 1968, p. 111–124.

GABRIEL, E.: Ölinseln im Persergolf.
G.R. 22 (1970), p. 309–315.

GABRIEL, E. (ed.): The Handbook of Dubai.
Dubai/Ahrensburg 1987.

GAUBE, H. AND E. WIRTH: Der Bazar von Isfahan.
Wiesbaden 1978. (Beihefte zum Tübinger Atlas des Vorderen Orients, Reihe B, no. 22).

GAUBE, H. AND E. WIRTH: Aleppo. Historische und geographische Beiträge zur baulichen Gestaltung, zur sozialen Organisation und zur wirtschaftlichen Dynamik einer vorderasiatischen Fernhandelsmetropole. 2 vols.
Wiesbaden 1984. (Beihefte zum Tübinger Atlas des Vorderen Orients, Reihe B, no. 58).

GEBHARDT, H.: Entwicklungsimpulse durch Verkehrserschliessung. Wirtschafts- und sozialgeographische Auswirkungen des jungen Infrastrukturausbaus in der Arabischen Republik Jemen. Wiesbaden 1987. (Jemen-Studien 6).

GEHRKE, U. AND H. MEHNER (ed.): Iran. Tübingen/Basel 1975. (Ländermonographien 5).

GIESE, E.: Sovchoz, Kolchoz und persönliche Nebenerwerbswirtschaft in Sowjet-Mittelasien. Münster 1973. (Westfälische Geogr. Studien 27).

GIESE, E.: Aufbau, Entwicklung und Genese der islamisch-orientalischen Stadt in Sowjet-Mittelasien.
Erdk. 34 (1980), p. 46–60.
Arabic translation in: E. Wirth (ed.): Deutsche Geographische Forschung im Orient. Erlangen/Beirut 1983, p. 444–465.

GIESE, E.: Nomaden in Kasachstan. Ihre Sesshaftwerdung und Einordnung in das Kolchos- und Sowchossystem.
G.R. 35 (1983), p. 575–588.

GIESSNER, K.: Naturgeographische Landschaftsanalyse der tunesischen Dorsale. Hannover 1964. (Jahrb. G.G. Hannover 1964).

GIESSNER, K.: Die jungquartäre Klimageschichte der Sahara – Stand und Vergleich der bisherigen Forschungsergebnisse.
In: K. Giessner and H.-G. Wagner (ed.): Geographische Probleme in Trockenräumen der Erde.
Würzburger Geogr. Arb. 53 (1981), p. 111–127.

GIESSNER, K.: Geography of Climates. Berlin/Stuttgart 1985. (Afrika Kartenwerk, B. N 5).

GRÖTZBACH, E.: Kulturgeographischer Wandel in Nordost-Afghanistan seit dem 19. Jahrhundert.
Meisenheim 1972. (Afghanische Studien 4).

GRÖTZBACH, E.: Formen bäuerlicher Wirtschaft an der Obergrenze der Dauersiedlung im Afghanischen Hindukusch.
In: C. Rathjens, C. Troll and H. Uhlig (ed.): Vergleichende Kulturgeographie der Hochgebirge des Südlichen Asien.
Erdwiss. Forschg. 5. Wiesbaden 1973, p. 52–61.

GRÖTZBACH, E.: Verstädterung und Städtebau in Afghanistan.
Afghanische Studien 14 (1976), p. 225–244.
Arabic translation in: E. Wirth (ed.): Deutsche Geographische Forschung im Orient. Erlangen/Beirut 1983, p. 466–480.

GRÖTZBACH, E.: Städte und Basare in Afghanistan. Eine stadtgeographische Untersuchung. Wiesbaden 1979. (Beihefte zum Tübinger Atlas des Vorderen Orients, Reihe B, no. 16).

GRÖTZBACH, E.: Zur Geographie des Erholungsverhaltens in einer traditionellen orientalischen Gesellschaft: Afghanistan.
In: E. Grötzbach (ed.): Freizeit und Erholung als Probleme der vergleichenden Kulturgeographie.
Eichstätter Beitr., Abt. Geogr. 1 (1981), p. 159–193.

GRÖTZBACH, E.: Yayla-Erholungsverkehr im östlichen Pontischen Gebirge (Nordost-Türkei). M.G.G. München 67 (1982), p. 91–124.

GROSSER, L.: Die Gramineen des Haraz (Arabische Republik Nordjemen) – Bestandsaufnahme, Standortansprüche und Nutzungsmöglichkeiten.
Diplom-Arb. Bayreuth 1985.

HAFEMANN, D.: Historical Geography. Berlin/Stuttgart 1981. (Afrika Kartenwerk, B. N 15).

HAGEDORN, H.: Untersuchungen über Relieftypen arider Räume an Beispielen aus dem Tibesti-Gebirge und seiner Umgebung.
Berlin 1971. (Z. f. Geomorph., N.F. Suppl. vol. 11).

HAGEDORN, H. et al: Pleistozäne Vergletscherungsspuren im Zentral-Iran.
Z. f. Geomorph., N.F. Suppl. vol. 23 (1975), p. 146–156.

HAHN, H.: Die Stadt Kabul (Afghanistan) und ihr Umland. Vol. 1: Gestaltwandel einer orientalischen Stadt.
Bonn 1964. (Bonner Geogr. Abh. 34).

HAHN, H.: Die Stadt Kabul (Afghanistan) und ihr Umland. Vol. 2: Sozialstruktur und wirtschaftliche Lage der Agrarbevölkerung im Stadtumland.
Bonn 1965. (Bonner Geogr. Abh. 35).

HAHN, H.: Ländliche Sozialstruktur und Entwicklungsmöglichkeiten in Afghanistan.
In: W. Lauer (ed.): Argumenta Geographica.
Coll. Geogr. 12 (1970), p. 271–285. (Troll-Festschrift).

HAHN, H.: Die traditionellen Lebensformen.
In: W. Kraus (ed.): Afghanistan. Teil 3: Mensch und Gesellschaft.
Tübingen/Basel 1972, p. 195–244.

HEINRITZ, G.: Grundbesitzstruktur und Bodenmarkt in Zypern. Eine sozialgeographische Untersuchung junger Entwicklungsprozesse.
Erlangen 1975. (Erlanger Geogr. Arb., Sonderband 2).

HEINRITZ, G.: Die Entwicklung junger Bewässerungsprojekte unter dem Einfluss gruppenspezifischen Pächterverhaltens.
G.Z. 65 (1977), p. 188–215.

HEINRITZ, G.: Zum Wanderungsverhalten der Westerner in der Republik Sudan. Beobachtungen im New Halfa- und im Rahad-Scheme.
Erdk. 35 (1981), p. 54–65.
Arabic translation in: E. Wirth (ed.): Deutsche Geographische Forschung im Orient. Erlangen/Beirut 1983, p. 365–385.

HEINRITZ, G.: Occupational Structures in Urban Squatter-Settlements of the Republic of Sudan.
GeoJournal 6 (1982), p. 41–47. [1982a].

HEINRITZ, G. (ed.): Problems of agricultural development in the Sudan. Selected papers of a seminar.
Göttingen 1982. (Forum 2). [1982b].

HERZOG, R.: Ethnography.
Berlin/Stuttgart 1981. (Afrika Kartenwerk, B. N 10).

HÖHFELD, V.: Anatolische Kleinstädte. Anlage, Verlegung und Wachstumsrichtung seit dem 19. Jahrhundert.
Erlangen 1977. (Erlanger Geogr. Arb., Sonderband 6).

HÖHFELD, V.: Städte und Städtewachstum im Vorderen Orient. Vergleichende Fallstudien zur regionalen Differenzierung jüngerer städtischer Entwicklungsprozesse im orientalisch-islamischen Kulturkreis. 2 vols.
Wiesbaden 1985. (Beihefte zum Tübinger Atlas des Vorderen Orients, Reihe B, no. 61).

HOPFINGER, H.: Geographische Aspekte des internationalen Handels mit gebrauchter Bekleidung. Perfektes Recycling oder Verlängerung asymmetrischer Handelsbeziehungen mit der Dritten Welt am Fallbeispiel Syrien.
Erdk. 39 (1985), p. 206–217.

HOSRY, M.: Sozialökonomische Auswirkungen der Agrarreform in Syrien.
Saarbrücken/Fort Lauderdale 1981. (Sozialökonomische Schriften zur Agrarentwicklung 43).

HÜMMER, PH.: Die Schädlingskatastrophe im Baumwollanbaugebiet der Çukurova/Türkei – Ihre geographischen, sozialen und wirtschaftlichen Konsequenzen.
Z. f. ausländische Landwirtschaft 16 (1977), p. 372–381.

HÜMMER, PH.: Siedlungsstrukturen und sozialräumliche Beziehungsmuster in der ländlichen Türkei. Aktionsräume altbäuerlicher und ehemals nomadischer Gruppen im Tertiärhügelland der Çukurova.
Bayreuth 1984. (Bayreuther Geowiss. Arb. 5).

HÜMMER, PH. AND M. SOYSAL: Investitionsverhalten ausländischer Arbeitnehmer in ihrem Heimatland. Die Türkei als Beispiel.
G.R. 31 (1979), p. 315–318.

HÜTTEROTH, W.-D.: Bergnomaden und Yaylabauern im mittleren kurdischen Taurus.
Marburg 1959. (Marburger Geogr. Schriften 11).

HÜTTEROTH, W.-D.: Getreidekonjunktur und jüngerer Siedlungsausbau im südlichen Inneranatolien.
Erdk. 16 (1962), p. 249–271.

HÜTTEROTH, W.-D.: Ländliche Siedlungen im südlichen Inneranatolien in den letzten vierhundert Jahren.
Göttingen 1968. (Göttinger Geogr. Abh. 46). [1968a].

HÜTTEROTH, W.-D.: Die Bedeutung kollektiver und individueller Landnahme für die Ausbildung von Streifen- und Blockfluren im Nahen Osten.
G.Z., Beihefte, Erdkundl. Wissen 18 (1968), p. 85–93. [1968b].

HÜTTEROTH, W.-D.: Schwankungen von Siedlungsdichte und Siedlungsgrenze in Palästina und Transjordanien seit dem 16. Jahrhundert.
Verh. 37. D.G. Kiel 1969.
Wiesbaden 1970, p. 463–475.
Arabic translation in: E. Wirth (ed.): Deutsche Geographische Forschung im Orient. Erlangen/Beirut 1983, p. 134–142.

HÜTTEROTH, W.-D.: Fragestellungen und Ergebnisse anthropogeographischer Forschung in Anatolien.
M.G.G. München 56 (1971), p. 77–94.

HÜTTEROTH, W.-D.: Zum Kenntnisstand über Verbreitung und Typen von Bergnomadismus und Halbnomadismus in den Gebirgs- und Plateaulandschaften Südwestasiens.
In: C. Rathjens, C. Troll and H. Uhlig (ed.): Vergleichende Kulturgeographie der Hochgebirge des südlichen Asien.
Erdwiss. Forschg. 5. Wiesbaden 1973, p. 145–156.
Arabic translation in: E. Wirth (ed.): Deutsche Geographische Forschung im Orient. Erlangen/Beirut 1983, p. 296–306.

HÜTTEROTH, W.-D.: The Influence of Social Structure on Land Division and Settlement in Inner Anatolia.
In: P. Benedict, E. Tümertekin and F. Mansur (ed.): Turkey. Geographic and social perspectives.
Leiden 1974, p. 19–47.

HÜTTEROTH, W.-D.: Die neuzeitliche Siedlungsexpansion in Steppe und Nomadenland im Orient.
In: H.-J. Nitz (ed.): Landerschliessung und Kulturlandschaftswandel an den Siedlungsgrenzen der Erde.
Göttinger Geogr. Abh. 66 (1976), p. 147–157.

HÜTTEROTH, W.-D.: Palästina und Transjordanien im 16. Jahrhundert. Wirtschaftsstruktur ländlicher Siedlungen nach osmanischen Steuerregistern.
Wiesbaden 1978. (Beihefte zum Tübinger Atlas des Vorderen Orients, Reihe B, no. 33).

HÜTTEROTH, W.-D.: Türkei.
Darmstadt 1982. (Wissenschaftliche Länderkunden 21).

HÜTTEROTH, W.-D.: Lebensformen im Orient: Nomaden – Bauern – Städter.
In: R. Kurzrock (ed.): Die Islamische Welt. Vol. 2.
Berlin 1984, p. 36–45. (Forschung und Information 36).

HÜTTEROTH, W.-D. AND K. ABDULFATTAH: Historical Geography of Palestine, Transjordan and Southern Syria in the Late 16th Century.
Erlangen 1977. (Erlanger Geogr. Arb., Sonderband 5).

IBRAHIM, F. N.: Das Handwerk in Tunesien – eine wirtschafts- und sozialgeographische Strukturanalyse.
Hannover 1975. (Jahrb. G.G. Hannover, Sonderheft 7).

IBRAHIM, F. N.: Desertification in Nord-Darfur. Untersuchungen zur Gefährdung des Naturpotentials durch nicht angepaßte Landnutzungsmethoden in der Sahelzone der Republik Sudan.
Hamburg 1980. (Hamburger Geogr. Studien 35).

IBRAHIM, F. N.: Anthropogeographische Ursachen der Desertifikation in den zentraltunesischen Steppen und der sudanischen Sahelzone.
In: E. Wirth (ed.): Deutsche Geographische Forschung im Orient. Erlangen/Beirut 1983, p. 80–103. [1983a]. (In Arabic).

IBRAHIM, F. N.: Das Handwerk in Tunesien aus sozialgeographischer Sicht.
In: E. Wirth (ed.): Deutsche Geographische Forschung im Orient. Erlangen/Beirut 1983, p. 481–503. [1983b]. (In Arabic).

IBRAHIM, F. N.: Ecological Imbalance in the Republic of the Sudan – with reference to desertification in Darfur.
Bayreuth 1984. (Bayreuther Geowiss. Arb. 6).

JÄKEL, D. et al: Forschungsstation Bardai. FU-Geologen in der Zentralsahara.
Berlin 1974. (FU Berlin, Pressedienst Wissenschaft 5, 1974).

JANZEN, J.: Landwirtschaftliche Aktiengesellschaften im Iran. Eine Fallstudie zur jüngeren Entwicklung der iranischen Agrarreform.
Saarbrücken 1976. (Sozialökonomische Schriften zur Agrarentwicklung 20).

JANZEN, J.: Die Nomaden Dhofars/Sultanat Oman. Traditionelle Lebensformen im Wandel.
Bamberg 1980. (Bamberger Geogr. Schriften 3).
Translated in English: Nomads in the Sultanate of Oman. Tradition and Development in Dhofar.
Boulder/London 1986.

JANZEN, J.: Nomadismus in Somalia. Struktur der Wanderweidewirtschaft und Hintergründe aktueller Entwicklungsprobleme im nomadischen Lebensraum – ein Überblick.
Afrika Spectrum 19 (1984), p. 149–171.

JANZEN, J.: Die ökonomische und ökologische Zukunft des Jubba-Tales/Süd-Somalia – Gedanken zur Problematik des geplanten Staudamm- und Bewässerungsprojektes aus geographischer Sicht.
Verh. 45. D.G. Berlin 1985. Wiesbaden 1987, p. 346–353.

JANZEN, J. AND F. SCHOLZ: Die Weihrauchwirtschaft Dhofars (Sultanat Oman). Eine wirtschaftsgeographische Studie zur Frage nach der Bedeutung einfacher Wirtschaftsresourcen in einem ölreichen Entwicklungsland.
Innsbrucker Geogr. Studien 5 (1979), p. 501–541.

JEBENS, A.: Wirtschafts- und sozialgeographische Untersuchung über das Heimgewerbe in Nordafghanistan . . .
Tübingen 1983. (Tübinger Geogr. Studien 87).

JENTSCH, CH.: Die Kareze in Afghanistan.
Erdk. 24 (1970), p. 112–120.

JENTSCH, CH.: Grundlagen und Möglichkeiten des Regenfeldbaus in Afghanistan.
Verh. 38. D.G. Erlangen-Nürnberg 1971. Wiesbaden 1972, p. 371–379.

JENTSCH, CH.: Das Nomadentum in Afghanistan.
Meisenheim 1973. (Afghan. Studien 9).

JENTSCH, CH.: Zur Höhengrenze der Siedlung in afghanischen Hochgebirgen.
Arb. aus dem Geogr. Inst. der Univ. des Saarlandes 29 (1980), p. 353–365.

JENTSCH, CH.: Die Nutzung der Hochgebirgsräume Afghanistans und ihre natur- und sozialräumlichen Grundlagen.
Frankfurter Wirtschafts- und Sozialgeogr. Schriften 46 (1981), p. 57–72.

JENTSCH, CH. AND R. LOOSE: Zur Geographie der ländlichen Siedlungen in Afghanistan.
Mannheim 1980. (Mannheimer Geogr. Arb. 6).

JUNGFER, E.: Das nordöstliche Djaz-Murian-Becken zwischen Bazman und Dalgan (Iran). Sein Nutzungspotential in Abhängigkeit von den hydrologischen Verhältnissen.
Erlangen 1978. (Erlanger Geogr. Arb., Sonderband 8).

JUNGFER, E.: Zur Frage der Grundwasserneubildung in Trockengebieten. Fallstudien aus der Arabischen Republik Jemen und dem Königreich Marokko.
Erlangen 1987. (Erlanger Geogr. Arb., Sonderband 18).

KLAER, W.: Eine Landnutzungskarte von Libanon.
Heidelberg/München 1962. (Heidelberger Geogr. Arb. 10). [1962a].

KLAER, W.: Untersuchungen zur klimagenetischen Geomorphologie in den Hochgebirgen Vorderasiens.
Heidelberg/München 1962. (Heidelberger Geogr. Arb. 11). [1962b].

KOCHWASSER, F. H. (ed.): Kuwait. Geschichte, Wesen und Funktion eines modernen arabischen Staates.
Tübingen/Basel 1969. (Ländermonographien 1).

KOPP, H.: Städte im östlichen iranischen Kaspitiefland. Ein Beitrag zur Kenntnis der jüngeren Entwicklung orientalischer Mittel- und Kleinstädte.
Erlangen 1973. (Erlanger Geogr. Arb. 33).

KOPP, H.: Der Einfluss temporärer Arbeitsemigration auf die Agrarentwicklung in der Arabischen Republik Jemen.
Erdk. 31 (1977), p. 226–230.
Arabic translation in: E. Wirth (ed.): Deutsche Geographische Forschung im Orient. Erlangen/Beirut 1983, p. 359–364. [1977a].

KOPP, H.: Al-Qāsim. Wirtschafts- und sozialgeographische Strukturen und Entwicklungsprozesse in einem Dorf des jemenitischen Hochlandes.
Wiesbaden 1977. (Beihefte zum Tübinger Atlas des Vorderen Orients, Reihe B, no. 31). [1977b].

KOPP, H.: Die Arabische Republik Jemen. Entwicklungsprobleme und Entwicklungsmöglichkeiten eines rohstoffarmen Entwicklungslandes.
G.R. 30 (1978), p. 88–93.
Arabic translation in: E. Wirth (ed.): Deutsche Geographische Forschung im Orient. Erlangen/Beirut 1983, p. 507–520.

KOPP, H.: Agrargeographie der Arabischen Republik Jemen. Landnutzung und agrarsoziale Verhältnisse in einem islamisch-orientalischen Entwicklungsland mit alter bäuerlicher Kultur.
Erlangen 1981. (Erlanger Geogr. Arb., Sonderband 11).

KOPP, H.: Die Entwicklung von Viehhaltung und Holznutzung im Rahmen der Agrarstruktur des Beckens von Aṭ Ṭūr.
Jemen-Studien 1 (1984), p. 75–86.

KOPP, H.: Land Usage and Its Implications on Yemeni Agriculture.
In: B.R. Pridham (ed.): Economy, society and culture in contemporary Yemen.
London 1985, p. 41–50.

KOPP, H. AND G. SCHWEIZER (ed.): Entwicklungsprozesse in der Arabischen Republik Jemen.
Wiesbaden 1984. (Jemen-Studien 1).

KORBY, W.: Probleme der industriellen Entwicklung und Konzentration in Iran.
Wiesbaden 1977. (Beihefte zum Tübinger Atlas des Vorderen Orients, Reihe B, no. 20).

KORTUM, G.: Die Marvdasht-Ebene in Fars. Grundlagen und Entwicklung einer alten iranischen Bewässerungslandschaft.
Kiel 1976. (Kieler Geogr. Schriften 44).

KRAUS, W. (ed.): Nomadismus als Entwicklungsproblem.
Bielefeld 1969. (Bochumer Schriften zur Entwicklungsforschung und Entwicklungspolitik 5).

KRAUS, W. (ed.): Afghanistan.
Tübingen/Basel 1972. (Ländermonographien 3).

KRAUSE, R. F.: Untersuchungen zur Bazarstruktur von Kairo.
Marburg 1985. (Marburger Geogr. Schriften 99).

KÜNDIG-STEINER, W. (ed.): Die Türkei. Raum und Mensch, Kultur und Wirtschaft in Gegenwart und Vergangenheit.
Tübingen/Basel 1974. (Ländermonographien 4).

LAUER, W. AND P. FRANKENBERG: Zur Klima- und Vegetationsgeschichte der westlichen Sahara.
Wiesbaden 1979. (Akad. d. Wiss. und d. Literatur. Mainz. Abh. d. Math.-Naturwiss. Kl. 1979, no. 1).

LEIDLMAIR, A.: Ḥaḍramaut. Bevölkerung und Wirtschaft im Wandel der Gegenwart.
Bonn 1961. (Bonner Geogr. Abh. 30).

LEIDLMAIR, A.: Umbruch und Bedeutungswandel im nomadischen Lebensraum des Orients.
G.Z. 53 (1965), p. 81–100.

LEIDLMAIR, A.: Die britischen Protektorate in Südarabien.
G.R. 18 (1966), p. 41–52.
Arabic translation in: E. Wirth (ed.): Deutsche Geographische Forschung im Orient. Erlangen/Beirut 1983, p. 112–125.

LEIPPERT, H. AND H. ZEIDLER: Vegetation Geography.
Berlin/Stuttgart 1984. (Afrika Kartenwerk, B. N 7).

MALIK, M. U. AND A. SCHIMMEL (ed.): Pakistan. Das Land und seine Menschen. Geschichte, Kultur, Staat und Wirtschaft.
Tübingen/Basel 1976. (Ländermonographien 6).

MAYER, J.: Lahore. Entwicklung und räumliche Ordnung seines zentralen Geschäftsbereichs.
Erlangen 1979. (Erlanger Geogr. Arb., Sonderband 9).

MECKELEIN, W.: Die Sahara-Forschungsfahrt 1954–55 der Gesellschaft für Erdkunde zu Berlin. Vorläufiger Expeditionsbericht.
Die Erde 7 (1955), p. 312–319.

MECKELEIN, W.: Forschungen in der zentralen Sahara. Geographische Ergebnisse der Sahara-Expedition 1954/55 der Gesellschaft für Erdkunde zu Berlin. Klimageomorphologie.
Braunschweig 1959.

MEDER, O.: Klimaökologie und Siedlungsgang auf dem Hochland von Iran in vor- und frühgeschichtlicher Zeit.
Marburg 1979. (Marburger Geogr. Schriften 80).

MENSCHING, H.: Marokko. Die Landschaften im Maghreb.
Heidelberg 1957. (Geogr. Handbücher).

MENSCHING, H.: Tunesien. Eine geographische Landeskunde.
Darmstadt 1968; 3. rev. ed. 1979. (Wissenschaftliche Länderkunden 1).

MENSCHING, H.: Nomadismus und Oasenwirtschaft im Maghreb.
Braunschweiger Geogr. Studien 3 (1971), p. 155–166.
Arabic translation in: E. Wirth (ed.): Deutsche Geographische Forschung im Orient. Erlangen/Beirut 1983, p. 307–316.

MENSCHING, H.: Der Mensch und seine Umwelt im Maghreb. Probleme und Gefahren der Inwertsetzung des Naturpotentials.
Hannover 1972.
Arabic translation in: E. Wirth (ed.): Deutsche Geographische Forschung im Orient. Erlangen/Beirut 1983, p. 73–79.

MENSCHING, H.: Desertifikation im zentraltunesischen Steppengebiet.
Nachrichten d. Akad. d. Wiss. in Göttingen, II. Math.-Phys. Kl. Folge 8.
Göttingen 1976, p. 99–118.

MENSCHING, H. AND E. WIRTH: Nordafrika und Vorderasien.
Frankfurt 1973, 2. rev. ed. 1980. (Fischer Länderkunde 4).

MESSERLI, B.: Das Problem der eiszeitlichen Vergletscherung am Libanon und Hermon.
Z. f. Geomorph., N.F. 10 (1966), p. 37–68.

MEYER, G.: Erschliessung und Entwicklung der ägyptischen Neulandgebiete.
Erdkunde 32 (1978), p. 212–227.
Arabic translation in: E. Wirth (ed.): Deutsche Geographische Forschung im Orient. Erlangen/Beirut 1983, p. 206–231.

MEYER, G.: Effects of the "New Valley" Project upon the development of the Egyptian oases.
Applied Geography and Development 15 (1980), p. 96–116.

MEYER, G.: Promotional measures and present state of development in the field of nomadism in Syria.
Applied Geography and Development 19 (1982), p. 97–107.

MEYER, G.: Ländliche Lebens- und Wirtschaftsformen Syriens im Wandel.
Sozialgeographische Studien zur Entwicklung im bäuerlichen und nomadischen Lebensraum.
Erlangen 1984. (Erlanger Geogr. Arb., Sonderband 16).

MEYER, G.: Labour emigration and internal migration in the Yemen Arab Republic – the urban building sector.
In: B. R. Pridham (ed.): Economy, society and culture in contemporary Yemen. London 1985, p. 147–164.

MEYER, G.: Arbeitsemigration, Binnenwanderung und Wirtschaftsentwicklung in der Arabischen Republik Jemen.
Wiesbaden 1986. (Jemen-Studien 2).

MEYER, G.: Waste-recycling as a livelihood in the informal sector – the example of refuse collectors in Cairo.
Applied Geography and Development 28 (1987).

MOMENI, M.: Malayer und sein Umland. Entwicklung, Struktur und Funktion einer Kleinstadt in Iran.
Marburg 1976. (Marburger Geogr. Schriften 68).

MÜLLER-HOHENSTEIN, K.: Nordafrikanische Trockensteppengesellschaften. Zur ökologischen Erklärung der räumlichen Differenzierung der Vegetation zwischen Mittelmeer und Sahara.
Erdk. 32 (1978), p. 28–39. [1978a].

MÜLLER-HOHENSTEIN, K.: Die ostmarokkanischen Hochplateaus. Ein Beitrag zur Regionalforschung und zur Biogeographie eines nordafrikanischen Trockensteppenraumes.
Erlangen 1978. (Erlanger Geogr. Arb., Sonderband 7). [1978b].

MÜLLER-HOHENSTEIN, K.: Das bilaterale Entwicklungshilfeprojekt DERRO im Rifgebirge Marokkos. Möglichkeiten und Grenzen der Entwicklung in einem Peripherraum.
In: G. Meyer (ed.): Geographische Aspekte der Entwicklungsländerproblematik. Reihe der Forschg. 8. Rheinfelden 1981, p. 54–60.
Arabic translation in: E. Wirth (ed.): Deutsche Geographische Forschung im Orient. Erlangen/Beirut 1983, p. 547–559.

MÜLLER-WIENER, W.: Burgen der Kreuzritter im Heiligen Land.
München/Berlin 1966.

MÜLLER-WIENER, W.: Bildlexikon zur Topographie Istanbuls.
Tübingen 1977.

MÜLLER-WIENER, W.: Der Bazar von Izmir. Studien zur Geschichte und Gestalt des Wirtschaftszentrums einer ägäischen Handelsmetropole.
M. Fränkische G. G. 27/28 (1980/81).
Erlangen 1982, p. 420–454.

PAPE, H.: Er Riad. Stadtgeographie und Stadtkartographie der Hauptstadt Saudi-Arabiens.
Paderborn 1977. (Bochumer Geogr. Arb., Sonderreihe 7).

PHILIPP, H.-J.: Geschichte und Entwicklung der Oase al-Hasa (Saudi-Arabien).
Vol. 1: Historischer Verlauf und traditionelles Bild.
Saarbrücken 1976. (Sozialökonomische Schriften zur Agrarentwicklung 23).

PHILIPP, H.-J.: Bewässerungsprojekte im Lichte der Sozialwissenschaften – Vorschläge für ihre Erforschung.
Z. f. Agrargeschichte und Agrarsoziologie 30 (1982), p. 66–78.

PHILIPP, H.-J.: Die landwirtschaftliche Modernisierung Saudi-Arabiens im 20. Jahrhundert.
Hamburg 1984. (Mitt. des Dt. Orient-Inst. 25).

PLANCK, U.: Die sozialen und ökonomischen Verhältnisse in einem iranischen Dorf.
Köln/Opladen 1962. (Forschungsber. des Landes Nordrhein-Westfalen, no. 1021). [1962a].

PLANCK, U.: Der Teilbau im Iran.
Z. f. ausländische Landwirtschaft 1 (1962), p. 47–81. [1962b].

PLANCK, U.: Rural Employment Problems in the United Arab Republic.
Employment Research Papers. Genf 1969.

PLANCK, U.: Die ländliche Türkei. Soziologie und Entwicklungstendenzen.
Frankfurt/M. 1972. (Z. f. ausländische Landwirtschaft, Materialsammlung 19).

PLANCK, U.: Iranische Dörfer nach der Bodenreform: Sozialorganisation und Sozialökonomik.
Opladen 1974. (Schriften des Dt. Orient-Inst., Materialien und Dokumente).

PLANCK, U.: Die Reintegrationsphase der iranischen Agrarreform.
Erdk. 29 (1975), p. 1–9.

PLANCK, U.: Die soziale Differenzierung der Landbevölkerung Irans infolge der Agrarreformen.
In: G. Schweizer (ed.): Interdisziplinäre Iran-Forschung.
Wiesbaden 1979, p. 43–58. (Beihefte zum Tübinger Atlas des Vorderen Orients, Reihe B, no. 40).
Arabic translation in: E. Wirth (ed.): Deutsche Geographische Forschung im Orient. Erlangen/Beirut 1983, p. 232–247.

POPP, H.: Entkolonialisierung und Agrarreform in Marokko. Das Beispiel Gharb.
Erdk. 34 (1980), p. 257–269. [1980a].

POPP, H.: Bodenreform und Genossenschaftswesen in Marokko.
Z. f. ausländische Landwirtschaft 19 (1980), p. 49–66.
Arabic translation in: E. Wirth (ed.): Deutsche Geographische Forschung im Orient. Erlangen/Beirut 1983, p. 263–277. [1980b].

POPP, H.: Bewässerungsprojekt Massa. Eine sozialgeographische Untersuchung der Planungsziele und der Anbaurealität in Südmarokko.
G.R. 34 (1982), p. 545–552.

POPP, H.: Moderne Bewässerungslandwirtschaft in Marokko. Staatliche und individuelle Entscheidungen in sozialgeographischer Sicht. 2 vols.
Erlangen 1983. (Erlanger Geogr. Arb., Sonderband 15). [1983a].

POPP, H.: Überpumpung als "manmade hazard" im Sousstal (Südmarokko). Sozialgeographische Aspekte exzessiver Grundwasserausbeutung.
Erdk. 37 (1983), p. 97–109. [1983b].

POPP, H.: Die mediterranen Küstenbereiche Nordmarokkos. Entwicklungsprobleme und staatlich gelenkte Entwicklungsprozesse in einer benachteiligten Region.
In: H. Popp and F. Tichy (ed.): Möglichkeiten, Grenzen und Schäden der Entwicklung in den Küstenräumen des Mittelmeergebietes.
Erlanger Geogr. Arb., Sonderband 17 (1985), p. 191–229.

RAPPENHÖNER, D.: Zum Futterwert der Gehölzpflanzen der Gebirgstihamah der Arabischen Republik Jemen.
Diplom-Arb. Bayreuth 1984.

RATHJENS, C.: Menschliche Einflüsse auf die Gestalt und Entwicklung der Tharr. Ein Beitrag zur Frage der anthropogenen Landschaftsentwicklung im Trockengebiet.
Arb. aus dem Geogr. Inst. der Univ. d. Saarlandes 4 (1959), p. 1–36.

RATHJENS, C.: Karawanenwege und Pässe im Kulturlandschaftswandel Afghanistans seit dem 19. Jahrhundert.
Tübingen 1962, p. 209–221. (Herrmann von Wissmann-Festschrift).

RATHJENS, C.: Menschliche Eingriffe in den Wasserhaushalt und ihre Bedeutung für die Geomorphologie der altweltlichen Trockengebiete (an Beispielen aus Afghanistan und dem nordwestlichen Indien).
Nova Acta Leopoldina, N.F. vol. 31, no. 176 (1966), p. 139–148.

RATHJENS, C.: Fragen der horizontalen und vertikalen Landschaftsgliederung im Hochgebirgssystem des Hindukusch.
In: C. Troll (ed.): Landschaftsökologie der Hochgebirge Eurasiens.
Erdwiss. Forschg. 4. Wiesbaden 1972, p. 205–220.

RATHJENS, C.: Witterungsbedingte Schwankungen der Ernährungsbasis in Afghanistan.
Erdk. 29 (1975), p. 182–188.
Arabic translation in: E. Wirth (ed.): Deutsche Geographische Forschung im Orient. Erlangen/Beirut 1983, p. 104–111.

RATHJENS, C.: Anliegen und Aufgaben deutscher geographischer Forschung in Afghanistan.
Orient 25 (1984), p. 551–560.

RICHTER, M. AND W. SCHMIEDECKEN: Das Oasenklima und sein ökologischer Stellenwert.
Erdk. 39 (1985), p. 179–197.

RIST, B.: Die Stadt Zābol – Zur wirtschaftlichen und sozialen Entwicklung einer Kleinstadt in Ost-Iran. (Sistān-Projekt I).
Marburg 1981. (Marburger Geogr. Schriften 86).

RITTER, W.: Der Erdölgolf. Struktur- und Entwicklungsprobleme der Länder am Arabisch-persischen Golf.
Köln 1983. (Problemräume der Welt 1).

RITTER, W.: Qatar. Ein arabisches Erdölemirat.
Nürnberg 1985. (Nürnberger wirtschafts- und sozialgeogr. Arb. 38).

RÖLLIG, W.: Tübinger Atlas des Vorderen Orients.
In: Ch. Schneider (ed.): Forschung in der Bundesrepublik Deutschland. Beispiele, Kritik, Vorschläge.
Weinheim 1983, p. 49–53. (Deutsche Forschungsgemeinschaft).

ROTHER, L.: Gedanken zur Stadtentwicklung in der Çukurova (Türkei). Von den Anfängen bis zur Mitte des 14. Jahrhunderts.
Wiesbaden 1972. (Beihefte zum Tübinger Atlas des Vorderen Orients, Reihe B, no. 3).

RUPPERT, H.: Beirut. Eine westlich geprägte Stadt des Orients.
Erlangen 1969. (Erlanger Geogr. Arb. 27).

SCHAMP, H.: Kharga. Von der Oasis magna zum Neuen Tal.
Die Erde 98 (1967), p. 173–202.

SCHAMP, H.: Sozialismus und Bodenreform in Ägypten.
Verh. 38. D.G. Erlangen-Nürnberg 1971. (Wiesbaden 1972), p. 278–288.
Arabic translation in: E. Wirth (ed.): Deutsche Geographische Forschung im Orient. Erlangen/Beirut 1983, p. 197–205.

SCHAMP, H. (ed.): Ägypten. Das alte Kulturland am Nil auf dem Weg in die Zukunft.
Tübingen/Basel 1977. (Ländermonographien 9). [1977a].

SCHAMP, H.: Ägypten. Entwicklungsprobleme eines alten Kulturlandes.
BP-Kurier 29, no. 1–2 (1977), p. 36–43.
Arabic translation in: E. Wirth (ed.): Deutsche Geographische Forschung im Orient. Erlangen/Beirut 1983, p. 521–532. [1977b].

SCHAMP, H.: Sadd el-Ali, der Hochdamm von Assuan – Fakten, Ziele, Konsequenzen.
Geowiss. in unserer Zeit 1 (1983), p. 51–59, 73–85.

SCHARABI, M.: Der Bazar. Das traditionelle Stadtzentrum im Nahen Osten und seine Handelseinrichtungen.
Tübingen 1985.

SCHLIEPHAKE, K.: Erdöl und regionale Entwicklung. Beispiele aus Algerien und Tunesien.
Hamburg 1975. (Hamburger Beitr. zur Afrika-Kunde 18).

SCHLIEPHAKE, K.: Städtische Wasserversorgung am Golf und im östlichen Saudi Arabien.
Geowiss. in unserer Zeit 1 (1983), p. 181–190.

SCHLIEPHAKE, K. (ed.): Tunesien.
Stuttgart 1984. (Ländermonographien 14).

SCHOLZ, F.: Belutschistan (Pakistan). Eine sozialgeographische Studie des Wandels in einem Nomadenland seit Beginn der Kolonialzeit.
Göttingen 1974. (Göttinger Geogr. Abh. 63). [1974a].

SCHOLZ, F.: Sesshaftmachung von Nomaden in der Upper Sind Frontier Province (Pakistan) im 19. Jahrhundert – Ein Beitrag zur Entwicklung und gegenwärtigen Situation einer peripheren Region in der Dritten Welt.
Geoforum 18 (1974), p. 29–46. [1974b].

SCHOLZ, F.: Sesshaftwerdung von Beduinen in Kuwait.
Erdk. 29 (1975), p. 223–234.

SCHOLZ, F.: Entwicklungstendenzen im Beduinentum der kleinen Staaten am Persischen/Arabischen Golf – Oman als Beispiel.
Mitt. der Österreichischen G.G. 118 (1976), p. 70–108.
Arabic translation in: E. Wirth (ed.): Deutsche Geographische Forschung im Orient. Erlangen/Beirut 1983, p. 317–341.

SCHOLZ, F.: Zur Entstehung von "Zentrum" und "Peripherie". Das omanische Entwicklungskonzept und Probleme bei seiner Realisierung.
In: K.-H. Hottes (ed.): Geographische Beiträge zur Entwicklungsländer-Forschung.
DGFK-Hefte 12. Bonn 1979, p. 123–144.
Arabic translation in: E. Wirth (ed.): Deutsche Geographische Forschung im Orient. Erlangen/Beirut 1983, p. 533–546.

SCHOLZ, F.: Sultanate of Oman. A geographical introduction to the country of Oman. Stuttgart 1980. [1980a].

SCHOLZ, F.: Teppichwirtschaft in Pakistan. Über die internen Probleme eines weltmarktabhängigen Produktionszweiges in einem Entwicklungsland. Die Erde 111 (1980), p. 301–327. [1980b].

SCHOLZ, F. (ed.): Beduinen im Zeichen des Erdöls. Wiesbaden 1981. (Beihefte zum Tübinger Atlas des Vorderen Orients, Reihe B, no. 45).

SCHOLZ, F.: Kupferproduktion in Oman – Ein Beispiel für die Bemühungen der kleinen arabischen Golfstaaten um Diversifizierung ihrer Wirtschaft. G.Z. 71 (1983), p. 229–246.

SCHOLZ, F.: Falaj-Oasen in Sharqiya, Inner-Oman. Die Erde 115 (1984), p. 273–294.

SCHOLZ, F. (ed.): Wirtschaftsmacht im Krisenherd. Die Golfstaaten. Braunschweig 1985. (Forum Erde), [1985a].

SCHOLZ, F. (ed.): Die kleinen Golfstaaten: Reichtum und Unterentwicklung – Ein Widerspruch? Stuttgart 1985. (Klett Länderprofile). [1985b].

SCHOLZ, F. AND J. JANZEN (ed.): Nomadismus – ein Entwicklungsproblem? Beiträge zu einem Nomadismus-Symposium, veranstaltet in der Gesellschaft für Erdkunde zu Berlin vom 11. bis 14. Februar 1982. Berlin 1982. (Abh. des Geogr. Inst., Anthropogeographie 33).

SCHWEIZER, G.: Nordost-Azerbaidschan und Shah Sevan-Nomaden. In: Strukturwandlungen im nomadisch-bäuerlichen Lebensraum des Orients. G.Z., Beihefte, Erdkundl.Wissen 26 (1970), p. 81–148.

SCHWEIZER, G.: Untersuchungen zur Physiogeographie von Ostanatolien und Nordwestiran. Tübingen 1975. (Tübinger Geogr. Studien 60).

SCHWEIZER, G. (ed.): Beiträge zur Geographie orientalischer Städte und Märkte. Wiesbaden 1977. (Beihefte zum Tübinger Atlas des Vorderen Orients, Reihe B, no. 24).

SCHWEIZER, G.: Bevölkerungsverteilung im Vorderen Orient. G.R. 30 (1978), p. 98–100. Arabic translation in: E. Wirth (ed.): Deutsche Geographische Forschung im Orient. Erlangen/Beirut 1983, p. 129–133.

SCHWEIZER, G. (ed.): Interdisziplinäre Iran-Forschung. Beiträge zur Kulturgeographie, Ethnologie, Soziologie und Neueren Geschichte. Wiesbaden 1979. (Beihefte zum Tübinger Atlas des Vorderen Orients, Reihe B, no. 40).

SCHWEIZER, G.: Gastarbeiter in Saudi-Arabien. Tübinger Geogr. Studien 80 (1980), p. 353–365. Arabic translation in: E. Wirth (ed.): Deutsche Geographische Forschung im Orient. Erlangen/Beirut 1983, p. 345–358.

SCHWEIZER, G.: The changing structure of Mountain Farming in the Pontus Mountains (Eastern Black Sea Region, Turkey). Applied Geography and Development 20 (1982), p. 46–62.

SCHWEIZER, G.: Social and economic change in the rural distribution system: Weekly markets in the Yemen Arab Republic. In: B.R. Pridham (ed.): Economy, Society and Culture in Contemporary Yemen. London 1985, p. 107–121.

SEGER, M.: Teheran. Eine stadtgeographische Studie. Wien/New York 1978.

ŞEN, E.: Die Entwicklung der Wohngebiete der Stadt Ankara. Ein Beitrag zum Gecekondu-Problem. G.Z. 60 (1972), p. 25–39.

SOYSAL, M.: Die Siedlungs- und Landschaftsentwicklung der Çukurova. Mit besonderer Berücksichtigung der Yüregir-Ebene.
Erlangen 1976. (Erlanger Geogr. Arb., Sonderband 4).

SPIEKER, U.: Libanesische Kleinstädte. Zentralörtliche Einrichtungen und ihre Inanspruchnahme in einem orientalischen Agrarraum.
Erlangen 1975. (Erlanger Geogr. Arb., Sonderband 3).

STADELBAUER, J.: Bahnbau und kulturgeographischer Wandel in Turkmenien. Einflüsse der Eisenbahn auf Raumstruktur, Wirtschaftsentwicklung und Verkehrsintegration in einem Grenzgebiet des russischen Machtbereichs.
Berlin (West) 1973. (Osteuropa-Inst. an der Freien Univ. Berlin. Wirtschaftswissenschaftliche Veröffentlichungen 34).

STADELBAUER, J.: Landwirtschaftliche Integration in den Subtropen der Sowjetunion. Überbetriebliche Zusammenarbeit im transkaukasischen Agrarraum.
Berlin 1983. (Giessener Abh. zur Agrar- und Wirtschaftsforschg. des Europäischen Ostens 120). [1983a].

STADELBAUER, J.: Studien zur Agrargeographie Transkaukasiens. Subtropische Landwirtschaft im gesamtsowjetischen Rahmen.
Berlin 1983. (Giessener Abh. zur Agrar- und Wirtschaftsforschg. des Europäischen Ostens 121). [1983b].

STADELBAUER, J.: Bergnomaden und Yaylabauern in Kaukasien. Zur demographischen Entwicklung und zum sozio-ökonomischen Wandel bei ethnischen Gruppen mit nicht-stationärer Tierhaltung.
Paideuma 30 (1984), p. 201–229.

STEWIG, R.: Bursa, Nordwestanatolien. Strukturwandel einer orientalischen Stadt unter dem Einfluss der Industrialisierung.
Kiel 1970. (Schriften des Geogr. Inst. der Univ. Kiel 32).

STEWIG, R.: Die Industrialisierung der Türkei.
Die Erde 103 (1972), p. 21–47.

STEWIG, R. et al: Bursa, Nordwestanatolien. Auswirkungen der Industrialisierung auf die Bevölkerungs- und Sozialstruktur . . .
Kiel 1980. (Kieler Geogr. Schriften 51).

STÖBER, G.: Die Afshār. Nomadismus im Raum Kermān/Zentraliran.
Marburg 1978. (Marburger Geogr. Schriften 76).

STÖBER, G.: Die Ṣayād-Fischer in Sistān. (Sistān-Projekt III).
Marburg 1981. (Marburger Geogr. Schriften 85).

STRÄTZ, C.: Standortansprüche und Vergesellschaftung der Gehölzvegetation des Haraz-Gebirges (Arabische Republik Jemen).
Diplom-Arb. Bayreuth 1985.

STRAUB, R.: Böden als Nutzungspotential im System der semiariden Tropen. Untersuchungen zur Bewertung von Bodeneigenschaften und Bodenwasserhaushalt am Beispiel des Beckens von Aṭ Ṭūr (Nordjemen).
Wiesbaden 1986. (Jemen-Studien 4).

STRUCK, E.: Landflucht in der Türkei. Die Auswirkungen im Herkunftsgebiet – dargestellt an einem Beispiel aus dem Übergangsraum von Inner- zu Ostanatolien (Provinz Sivas).
Passau 1984. (Passauer Schriften zur Geogr. 1).

STUCKMANN, G.: Hydrogeographische Untersuchungen im Bereich der mittleren Medjerda und ihre Bedeutung für den Landschaftshaushalt in Nordtunesien.
Hannover 1968. (Jahrb. G.G. Hannover, Sonderheft 3).

TOEPFER, H.: Wirtschafts- und sozialgeographische Fallstudien in ländlichen Gebieten Afghanistans.
Bonn 1972. (Bonner Geogr. Abh. 46).

WAGNER, H.-G.: Das Siedlungsgefüge im südlichen Ostalgerien (Nememcha).
Erdk. 25 (1971), p. 118–135.
Arabic translation in: E. Wirth (ed.): Deutsche Geographische Forschung im Orient. Erlangen/Beirut 1983, p. 143–163.

WAGNER, H.-G.: Die Souks in der Medina von Tunis. Versuch einer Standortanalyse von Einzelhandel und Handwerk in einer nordafrikanischen Stadt.
In: R. Stewig and H.-G. Wagner (ed.): Kulturgeographische Untersuchungen im islamischen Orient.
Schriften des Geogr. Inst. d. Univ. Kiel 38 (1973), p. 91–142.

WAGNER, H.-G.: Population Geography.
Berlin/Stuttgart 1981. (Afrika Kartenwerk, B. N 8).

WAGNER, H.-G.: Settlement Geography.
Berlin/Stuttgart 1983. (Afrika Kartenwerk, B. N 9).

WAGNER, H.-G.: Die Entwicklung der räumlichen Bevölkerungsverteilung.
In: K. Schliephake (ed.): Tunesien.
Stuttgart 1984, p. 232–263. (Ländermonographien 14).

WIEBE, D.: Stadtstruktur und kulturgeographischer Wandel in Kandahar und Südafghanistan.
Kiel 1978. (Kieler Geogr. Schriften 48).

WIEBE, D.: Afghanistan. Ein mittelasiatisches Entwicklungsland im Umbruch.
Stuttgart 1984. (Klett Länderprofile).

WIRTH, E.: Die Lehmhüttensiedlungen der Stadt Bagdad. Ein Beitrag zur Sozialgeographie orientalischer Städte.
Erdk. 8 (1954), p. 309–316.

WIRTH, E.: Der heutige Irak als Beispiel orientalischen Wirtschaftsgeistes.
Die Erde 8 (1956), p. 30–50.

WIRTH, E.: Agrargeographie des Irak.
Hamburg 1962. (Hamburger Geogr. Studien 13).

WIRTH, E.: Die Ackerebenen Nordostsyriens.
G.Z. 52 (1964), p. 7–42.

WIRTH, E.: Strukturwandlungen und Entwicklungstendenzen der orientalischen Stadt. Versuch eines Überblicks.
Erdk. 22 (1968), p. 101–128.

WIRTH, E.: Syrien. Eine geographische Landeskunde.
Darmstadt 1971. (Wissenschaftliche Länderkunden 4/5).

WIRTH, E.: Die Beziehungen der orientalisch-islamischen Stadt zum umgebenden Lande. Ein Beitrag zur Theorie des Rentenkapitalismus.
In: E. Meynen (ed.): Geographie heute. Einheit und Vielfalt. Ernst Plewe zu seinem 65. Geburtstag.
G.Z., Beihefte, Erdkundl. Wissen 33 (1973), p. 323–333.

WIRTH, E.: Zum Problem des Bazars (sūq, çarşı). Versuch einer Begriffsbestimmung und Theorie des traditionellen Wirtschaftszentrums der orientalisch-islamischen Stadt.
Der Islam 51 (1974), p. 203–260; 52 (1975), p. 6–46. [1974/75].

WIRTH, E.: Die orientalische Stadt. Ein Überblick aufgrund jüngerer Forschungen zur materiellen Kultur.
Saeculum 26 (1975), p. 45–94.

WIRTH, E.: Der Orientteppich und Europa. Ein Beitrag zu den vielfältigen Aspekten west-östlicher Kulturkontakte und Wirtschaftsbeziehungen.
Erlangen 1976. (Erlanger Geogr. Arb. 37).

WIRTH, E.: Orientalistik und Orientforschung. Aufgaben und Probleme aus der Sicht der Nachbarwissenschaften.
19. Dt. Orientalistentag Freiburg 1975. Z. der Dt. Morgenländischen Gesellschaft, Supplement Band III, 1. Wiesbaden 1977, p. LV-LXXXII. [1977a].

WIRTH, E.: Technological innovation and development in the Middle East. The impact of pump irrigation for the modernization of agriculture and the resulting change[s] of land use and settlement [patterns].
In: R.C. Eidt et al. (ed.): Man, Culture, and Settlement. Festschrift to Prof. R. L. Singh.
New Delhi 1977, p. 396–403. (National Geographical Society of India, Research Publication 17). [1977b].

WIRTH, E.: Agrarreform und ländliche Abwanderung im Irak. Sozialer Wandel und wirtschaftliche Dynamik in einer "sozialistischen Gesellschaft" mit Freiräumen für Privatinitiative.
Erdk. 36 (1982), p. 188–198.

WIRTH, E. (ed.): Deutsche geographische Forschung im Orient. Ein Überblick anhand ausgewählter gegenwartsbezogener Beiträge zur Geographie des Menschen.
German geographical research in the Middle East. A survey on the basis of selected articles on human geography which are of current relevance.
Erlangen/Beirut 1983. (Erlanger Geogr. Arb., Sonderband 14).

WIRTH, E.: Les forces économiques et sociales de l'aménagement de Sana'a. Transformations de la médina et croissance périphérique d'une ville d'aspect médiéval.
In: F. Métral (ed.): Politiques urbaines dans le Monde Arabe. Table Ronde C.N.R.S. tenue à Lyon . . . 1982.
Lyon 1984, p. 451–460. (Etudes sur le Monde Arabe 1).

WIRTH, E.: Tradition und Innovation im Handwerk und Kleingewerbe der vorderorientalischen Stadt. Strukturwandlungen und Überlebensstrategien in den vergangenen 150 Jahren.
Die Welt des Islams, Neue ser. 25 (1985), p. 174–222.

WIRTH, E.: Aleppo im 19. Jahrhundert. Ein Beispiel für Stabilität und Dynamik spätosmanischer Wirtschaft.
Erlangen 1986. Revisted print from: H. G. Majer (ed.): Osmanistische Studien zur Wirtschafts- und Sozialgeschichte. Wiesbaden 1986.

ZIA TAVANA, M. H.: Die Agrarlandschaft Irānisch-Sistāns. (Sistān-Projekt II).
Marburg 1983. (Marburger Geogr. Schriften 91).

ZIMPEL, H. G.: Bevölkerungsdichte und Siedlungsverteilung im Bereich der mediterranen Randschwelle der Arabischen Halbinsel.
M.G.G. München 50 (1965), p. 47–75.

ZIMPEL, H. G.: Niederschlagsvariabilität und Anbau-Unsicherheit in Syrien.
M.G.G. München 54 (1969), p. 89–114.

C. *Cartography and mapping programs*

Afrika-Kartenwerk, sheet Tunis:

Sheet N 1: Topography (G. STUCKMANN 1977).
Sheet N 2: Relief and Landforms (K. GIESSNER 1982).
Sheet N 4: Pedology (R. BRECHTEL and W. ROHMER 1978).
Sheet N 5: Hygric and thermic climatic types (K. GIESSNER 1983).

Sheet N 6: Surface water and groundwater (G. STUCKMANN and K. GIESSNER 1983).
Sheet N 7: Vegetation units (H. LEIPPERT and H. ZEIDLER 1983).
Sheet N 8: Population 1966 (H.-G. WAGNER 1982).
Sheet N 9: Settlements (H.-G. WAGNER 1982).
Sheet N 10: Ethnography (R. BECKER and R. HERZOG 1976).
Sheet N 11: Land use (H. ACHENBACH 1976).
Sheet N 12: Mining – Manufacturing Industries – Handicrafts – Fishing – Energy – Tourism (A. ARNOLD 1976).
Sheet N 13: Traffic 1966/1972 (A. ARNOLD 1981).
Sheet N 15: Roman Cultural Landscape at about 300 A.D. (D. HAFEMANN 1977).

Tübinger Atlas des Vorderen Orients:

1. Lieferung 1977:	A VIII 2	– Middle East. Population Distribution
	A IX 9,4	– Isfahan/Eşfahān Bazar (Iran)
2. Lieferung 1978:	A VIII 4	– Middle East. Urban and Rural Population
3. Lieferung 1979:	A VIII 7	– Lebanon. Religions
4. Lieferung 1980:	A III 5	– Central and Southern Arabia. Geomorphology
5. Lieferung 1981:	A IX 7	– Afghanistan. Towns and market places at the end of the 19th century
	A IX 8	– Afghanistan. Towns and market places round 1973
6. Lieferung 1982:	A IX 6	– Middle East. Towns and Urban Growth
	A IX 10	– Examples of Small and Medium-sized Oriental Cities. Urban Structure (10.1–10.4)
7. Lieferung 1983:	A III 4	– Levant. Geomorphology
8. Lieferung 1984:	A II 3	– Middle East. Seismotectonics, Historical Earthquakes
	A X 18	– Middle East. Traffic (Development of Land and Sea Traffic Routes)
9. Lieferung 1984:	A IX 10	– Examples of Small and Medium-sizend Oriental Cities. Urban Structure (10.5–10.7)
10. Lieferung 1985:	A I 4	– Mesopotamia. Relief, Hydrology and Settlements
	A X 10	– Land Use. Examples
11. Lieferung 1985:	A IV 5	– Middle East, Rainfall Reliability
	A IV 6	– Middle East. Seasonal Rainfall Patterns
	A VII 4	– Levant. Natural Regions
	A X 4	– Mesopotamia. Land Utilization
12. Lieferung 1986:	A VIII 5	– Middle East. Population Mobility
13. Lieferung 1986:	A V 1	– Middle East. Hydrogeography
	A IX 9	– Examples of Large Oriental Cities, San'a (Yemen). Urban Structure
	A X 12	– Pastoral Migration Systems. Examples Gab (Syria)
	B IX 15	– Middle East. The Economy in the 19th century

German geographical research in the High Mountains of the World

by

Carl Rathjens (Saarbrücken)

1. Development before the Last War

Only a relatively small section of the Alps is situated inside Germany but it represents a typical example of high mountains*, where since the 18th century intensive research work has been carried out. For more than 100 years Geography has contributed considerably and increasingly to this research. It is only too natural that a close co-operation arose with the other Alpine countries, especially with German speaking Austria and Switzerland. With Austria in particular there has always been a stimulating exchange of ideas and of scientists, and Alpinism and its scientific activities were organized jointly till the Second World War in the "*Deutscher und Österreichischer Alpenverein*", which considered as its important task not only the touristic opening of the Alps but also their scientific exploration and that of distant high mountains in subtropical and tropical regions. Relatively independent of this German-Austrian co-operation was the development in Switzerland, the more so as geographical high mountain research there took second place to geology, botany and ethnology for a longer time.

In Germany geographic research of high mountains reached a first peak after the First World War. It was mainly the influence of *Albrecht Penck* (1858–1945), the great German geographer and researcher of the ice age who together with *Eduard Brückner* established in the Alps the fourfold division of the Quaternary ice ages and who established together with William Morris Davis the first consistent concept of geomorphology. Accordingly topics of geomorphology, especially of glacial geomorphology have for a long time been in the forefront of interest of German geographers.

---

* As the English language does not provide an altogether synonymous term for the concept of "Hochgebirge", the following treatise will refer to this using the term of "high mountains". E. W.

The tectonically controlled vertical arrangement (Stockwerkbau) of mountains, the summit levels (Gipfelflur), relics of older planation surfaces, valley floors and piedmonts or foot slopes (Fussflächen) were treated in a great number of papers besides the glacial features. Only gradually did the researchers direct their attention to the features of solifluction in the periglacial zone of the mountains. The state of knowledge in this field was reviewed by Fritz MACHATSCHEK (1876–1957) in his great work on the relief of the earth (1938/40, 2. ed. 1955), the first comprehensive regional geomorphology.

Carl TROLL, without doubt the most outstanding German researcher of the geography of high mountains in the last decades, also was inspired primarily by Albrecht Penck. In 1924 Troll produced the first detailed investigation of Pleistocene foreland glaciation, with the geomorphological mapping at 1:100.000 of the Inn-Chiemsee glacier at the northern margin of the Alps. His profound knowledge of botany and vegetation geography, his research work in the South American Andes, in Eastern Africa and as a member of one of the German Nanga Parbat expeditions in the Himalayas enabled him to develop a concept of high mountains as a geoecological system in a three-dimensional and landscape oriented (landschaftskundlich) view. For this purpose he used the climate and vegetation of the mountains as indicators to which he dedicated numerous special studies. Already in 1941 Troll compared high mountains of different climatic zones of the world and elaborated on their essential differences. His geographical definition of the nature of high mountains has been valid until today, even though it had to be modified and extended.

After the war Carl Troll assumed a leading role in German geography: he was president of the International Geographical Union from 1960 to 1964, and in 1968 he founded the IGU-Commission "High Altitude Geoecology" over which he presided until 1972. His travels led him to many other high mountains of the world, and up to his death in 1975 his scientific activity was indefatigable. The last conference in which he participated was the German geographical biennial conference (Geographentag) in Innsbruck/Austria in 1975, at which high mountains were the major topics of papers and discussions. His extensive life work and his numerous disciples will ensure that his concept will persist into the future.

The Alpine Mountaineering Associations (Alpenvereine) in the Eastern Alps have from the beginning of their activities stressed the value of good maps on a large scale. Since such maps were not available from the official survey like in Switzerland, their own methods of cartographical mapping and production had to be developed. The so called "Alpenvereinskarten" are exemplary even today as outstanding high mountain maps. They are an important basis of research for geography and neighbouring disciplines like geology or glaciology, and also for some human sciences. The cartographer Richard FINSTERWALDER (1899–1963) stimulated terrestrial photogrammetry and glaciology and co-operated closely with geography.

In this way the first reliable *topographic maps for foreign high mountains* were produced: the Fedshenko glacier in the Alai Pamir region in the Soviet Union, the Cordillera Real in Bolivia and the Cordillera Blanca in Peru, both ranges in the South American Andes, the Zemu glacier near the Kangchendzenga and the Nanga Parbat massif in the Himalayas. Geologists, glaciologists and vegetation geographers in particular were able to base their investigations on such maps. At the Rakhiot glacier near the Nanga Parbat for instance the block movement of a subtropical high mountain glacier was observed for the first time. In 1939 Carl Troll produced a vegetation map of the whole mountain massif demonstrating the vertical zonation of vegetation from the desert Indus valley up to the snow line at an altitude of more than 5000 m.

In contrast to this clear predominance of topics of physical geography, *Human Geography* was little involved in the German high mountain research between the wars. Human Geography showed some interest in the economy of Alpine pastures to which Robert SIEGER directed attention as early as 1907, in the frontiers and locations of settlements, but it was less interested in the agricultural utilization of mountains. However, more research effort was concentrated on the Alps when 1928 Norbert KREBS published his book about the "Eastern Alps and modern Austria" and thus produced a long-lasting model of a descriptive treatment in regional geography. The dissertation of Hermann v. WISSMANN in 1927 dealing with the recession of the mountain peasantry in the Mitter Enns valley also showed a new direction. In 1931 Carl TROLL investigated the geographical basis of the Andine Indian cultures and of the Inca empire. In 1935 and 1943 respectively Wilhelm CREDNER and Hermann v. WISSMANN studied the ethnical stratification in the high mountains of Southeast Asia, Thailand and Southern China. There was a good foundation and methodology from which German geography was able to resume work after the forced interval of the war and the first post war years.

2. Regional focal points since 1945

When the German geographical research in high mountains began to recover gradually after 1945, at first its interests concentrated on the Alps, mainly on the *Eastern Alps* in Bavaria and Austria. The reason for this was spatial proximity and a common language, as well as the availability of excellent large scale topographic and geological maps, of meteorological data and of abundant statistical material. Thus the research work was more and more intensified. In the publications of the Alpine Associations (Alpenvereine) and of the geographical departments of universities in and near the Alps numerous important results of research were produced. Less German research work was done in Switzerland where geography of high mountains went its own independent way after 1945 and developed its own objectives and working methods, and the same applies to the French, Italian and Jugoslavian parts of the Alps.

Soon after the war high mountains of the *Mediterranean region* were included in the investigations. Much of this research dealt with special questions such as the trend of the Pleistocene snow line and the lower margin of glaciations (Sierra Nevada in Spain, mountains of the Maghreb, Anatolia, Lebanon and others), the timber line or certain formations of vegetation and Karst features which have their most pronounced forms in the Dinarian mountains. Human geography became preoccupied with the phenomena of the transhumance or the depopulation of Mediterranean mountains.

The renewed interest for tropical and subtropical high mountains took on great importance. From the tradition of the prewar era two regions of research persisted, the Himalayas and the South American Andes. In the *Himalayas* some expeditions in which geographers participated investigated the massif of Nanga Parbat and the Karakorum mountains. Since 1951 the Nepal research group supported by the Thyssen Foundation, which later on expanded into the Association for Comparative High Mountain Research in Munich, has attracted most energy. Within this institution excellent interdisciplinary cooperation was practised and it can look back upon a proud series of scientific publications (Khumbu Himal). It also produced some large scale topographic maps of the Nepal Himalayas which are among the best results of cartography of high mountains in the world.

No comparable research organization was founded for the *Andes* as similar plans by Carl Troll did not come to fruition. However despite this, many individual research expeditions brought together an immense amount of material from the landbridge of Middle America down to Patagonia. It has become so extensive that it is difficult to be grasped by one geographer. It was published in the volumes of “Erdwissenschaftliche Forschung” of the Academy of Sciences and Literature in Mainz, in the “Bonner Geographische Abhandlungen” and in many other publications. Within the interdisciplinary Mexico-Project which was coordinated by geographers and supported by the “Deutsche Forschungsgemeinschaft” (German Research Foundation) geographical research work was undertaken on some high volcanoes.

In addition to these centres some new regions were researched which had not been accessible or were difficult to reach before the war. Unfortunately some of these are no longer accessible today especially in the Near East (Eastern Anatolia, Iran, Lebanon, Hindukush in Afghanistan). In Africa all high mountains from Morocco and Tenerife to Ethiopia and Eastern Africa were visited by German geographers. An example is the study of Peter HÖLLERMANN (1982) about morphodynamics and geoecology of some Canary Islands. Much less German research work has of course been undertaken in the well investigated mountains of the USA and Japan. We mention here the study of Ingrid HENNING (1974) about the geoecology of the islands of Hawaii. For political reasons the high mountains of Central, East and Southeast Asia remained closed for a long time; only Southwestern China and the highlands of Tibet are now beginning to be opened up to a few European and American researchers. All in all the world of high mountains is very unevenly covered by the German speaking literature and this is clearly expressed by the lack of a manual or handbook on high mountain geography. Carl Troll died before he could accomplish such a book.

3. Climatically controlled features

The exceptional position of mountains with respect to climate has to be seen in close connection with the *vertical zonation of geomorphological and vegetational features.* Excepting certain air circulations which are conditioned by relief (phenomena of foehn, wind systems in valleys) the zonation corresponds to the vertical or hypsometric change in the geographical features (Formenwandel) in the system of landscapes by Hermann LAUTENSACH (1952). Decrease in temperature and increase in precipitation are the decisive factors, but precise statements are complicated by the lack of meteorological stations at high altitudes.

Some authors have been preoccupied with the role of temperature in the vertical zonation of solifluction and the snow line, in many cases stimulated by the occurrence of permafrost which have also been increasingly observed in high mountains of middle and lower latitudes. Wolfgang WEISCHET (1965) and Wilhelm LAUER (1975) have elaborated on the differences between the nontropical advective and the tropical convective type of vertical distribution of precipitation: in the mountains outside the tropics precipitation increases directly with increasing altitude, while in the tropics the zone of maximum precipitation is in the foothills or the periphery of the mountains. Penitentes snow (Büsserschnee) was recognized as a symptom of summer-dry subtropical high mountains (Carl TROLL 1942, Günther SCHWEIZER 1969).

The foehn has been a long-standing topic of geographical and meteorological research in European mountains. Its worldwide distribution (Chinook, Puelche, Zonda etc.) was recognized early. Carl TROLL (1952) observed for the first time the peculiar nature of the local winds in tropical mountains and described them so impressively that today we speak of the Troll-effect. After the fundamental investigations of Carl TROLL in the Andes Ulrich SCHWEINFURTH (1957) produced a vegetation map of the Himalayas and was able to prove that the aridity of the big valleys is conditioned by air circulation.

4. Geomorphological forms and processes

Geomorphological investigations have remained in the centre of geographical research for a long time and only recently have they been overtaken by the more complex questions of geoecology. The role of high mountains in the *climatic change of the Quaternary*, the height and oscillations of the snow line and of the lower limit of solifluction, extensions of glaciation and types of movement of glaciers have been important themes. In this way essential knowledge about changes of climate and landscapes in the Quaternary of the whole world was gained.

But with the refinement of methods and with more detailed results it has become apparent that certain ideas about the shift of the climatic zones, about Pleistocene depressions of the snow line and about the simultaneousness of certain stages of glaciations were premature. Today we are no longer certain that the maximum glaciations of mountains all over the globe occured contemporaneously. Especially for subtropical mountains some concepts had to be corrected (Klaus HEINE 1977, Helmut STINGL and Karsten GARLEFF 1978). It must, however, be stated that the hypothesis of extremely low and extensive glaciations in high mountain ranges of Iran (Matthias KUHLE 1976) are highly improbable as are high rates of depression of the lower limit of solifluction in arid climates (Jürgen HÖVERMANN 1960, Carl RATHJENS 1965, Peter HÖLLERMANN 1972).

Besides the research about the ice ages and former glaciations more and more investigations examined geomorphological processes. Since Carl TROLL's pioneering work (1944, 1947) special attention has been given to the *vertical zonation of solifluction*, also known as subnival or periglacial zone, and its specific processes caused by alternating frost and thaw. The vertical extent of this zone, its processes and features are an essential contribution to the regional geography of high mountains. On this aspect Hans POSER (last 1977) conducted or stimulated numerous investigations into mountains between the Arctic and the equator. Prominence has to be given to the studies of Peter HÖLLERMANN (1982) about recent morphodynamics on Tenerife. Work in the tropical high mountains of New Guinea is represented by Ernst LÖFFLER (1975). Matthias KUHLE (1982), who worked in the extremely rugged high mountains of the Himalayas and in Tibet, has referred to the existence of a "pergelid" zone without frost change and solifluction.

For some time the problem of the *rock glaciers* has been of special interest. It was first investigated in the Alps by Wolfgang PILLEWITZER (1957) and Dietrich BARSCH (1969), in the Near East by Wendelin KLAER (1962) and Günther SCHWEIZER (1972), in the Hindukush by Erwin GRÖTZBACH (1965), and more recently by Peter HÖLLERMANN (1983) in a comparison between European and North American mountains. Without doubt rock glaciers are a phenomenon of the vertical zone of solifluction in a semihumid or continental climate with permafrost, but the discussion is going on.

More studies were undertaken about mass-movements in steep relief occurring often as catastrophic events. Gerhard ABELE (1974) investigated the rockfalls and landslides of late and postglacial times in the Alps. Other authors have studied the debris transport by avalanches or mud flows after earthquakes or volcanic eruptions in many mountains of the world. As a young branch of Applied Geomorphology mapping of hazards in high mountains has developed out of this. The importance of actual morphodynamics to the geoecological pattern of high mountains has been recognized for some time, but in this field geomorphology has been lagging behind questions about the edaphic conditions of vegetation and soils.

5. Vegetation geography and geoecology

With reference to *Alexander von Humboldt* and his research work in Latin America *Carl Troll* already had made the vertical zonation of vegetation the focus of geographical high mountain research and demonstrated this with his studies of the Andes of Bolivia and Peru, the East African volcanoes, the Northwestern Himalayas and Sikkim. His vegetation map of the Nanga Parbat (1939) stimulated a great number of further work using and clarifying many terms which had been coined by Carl Troll. Worth mentioning are the investigations of Ulrich SCHWEINFURTH in the Himalayas (1957) and in New Zealand (1966) and of Wilhelm LAUER in Central America (1959), Mexico (1973) and in the Northern Andes (1979) for which the term "cold tropics" was also introduced.

It would lead too far to enumerate here all the members of the different research projects and study groups in the mountains around the earth, the more so, as they were not always geographers but often botanists. Botanical mapping, too, has become most valuable to the vegetation geography of high mountains. The vegetation surveys in Iran, in the Hindukush of Afghanistan and in the Nepal Himalayas deserve to be mentioned here as do the investigations of Friedrich-Karl HOLTMEIER (1967, 1978) on the upper timberline in the Alps, in Scandinavia and in North America.

Geoecological questions should consider climate, relief, morphodynamics, soils, vegetation and fauna in the same way, and in many cases also utilization by man. Although the necessary measurements and data often do not exist in a sufficient quantity, vegetation may be an easily recognizable indicator of other geoecological factors. Therefore maps called "ecological" often are in effect maps of vegetation. But there are some remarkable examples of geoecological investigations in German geography of high mountains, as for instance by Willibald HAFFNER (1979) of Nepal or the already mentioned Ingrid HENNING (1974) of Hawaii and Peter HÖLLERMANN (1982) of Tenerife. In spite of many necessary qualifications and objections it can be said today that a system of high mountain research is emerging in physical geography (Carl RATHJENS 1982) based on the viewpoints of landscape ecology.

6. Human geography

This last statement about the development of physical geography in high mountain research cannot yet be applied to human geography. Its investigations are widely scattered in the choice of themes and regions of work and have not found a homogeneous and generally accepted concept. First attempts in this direction have been made by Erwin GRÖTZBACH (since 1975) and Harald UHLIG (1984).

Human geography has a distinct centre of interest in the *Alps* where special attention is directed to high pasture economy and to land use at the upper limit of agriculture. Karl RUPPERT (since 1963) and his team have produced many studies of the geography of high pastures (Almgeographie), partly in co-operation with Jugoslav researchers. Felix MONHEIM (1954) studied the agricultural geography of the western high Alps and introduced the term "high altitude fallow" (Höhenbrache) at the upper limit of agriculture. He transferred his findings successfully to the high Andes of South America.

The *economy of high pastures* belongs to the larger field of grazing and of seasonal migrations of man and cattle to high mountains. Therefore these facts have met with a growing interest in many mountains of the world, even though their importance for tropical high mountains is not yet fully understood. The Old World countries with old civilisations have developed the most complicated systems and have therefore received the greatest attention. Arnold BEUERMANN (1967) investigated the transhumance and high pasture economies in the mountains of Southeastern Europe, Wolf-Dieter HÜTTEROTH (1959) the mountain nomads and yayla peasants in the Taurus of Kurdistan, Günther SCHWEIZER (1970) a group of nomads in Northwestern Iran, Christoph JENTSCH (1973) the organization of nomadism in Afghanistan, Harald UHLIG (since1963) migrating herdsmen and altitudinal zones of agriculture in Kashmir. Later on these studies were extended over the whole range of the Himalayas and into Southeast Asia. In several studies Gisbert RINSCHEDE (1984) undertook a comparison of the old transhumance in the western Mediterranean region with the young migrating-cattle economy in the high mountains of the Western USA.

Many papers about *rural landscapes* in mountainous countries concern areas all over the world, but they have not yet been combined into a conclusive system. The upper limits of cultivation for barley, potatoes, and rice can already be pursued on a worldwide scale. The rice cultivation systems dominant in the Near, Middle and Far East have been studied by Harald UHLIG (since 1973); cultivation in common field systems (Zelgen) is used in many mountains of the Old World. But investigations on types of houses and settlements are still at their beginning. Because the traditional forms of land use and rural settlement in high mountains are undergoing rapid changes today, further investigations are urgently required. The book of Christian KLEINERT (1983) about settlement and natural environment in the Nepal Himalayas is an excellent example. Co-operation with ethnology is beginning to develop well in this field.

The age of industrialization brought with it many changes which affected even the remotest valleys in high mountains. In many mountain areas the pressures of increasing population and economic hardship have resulted in *emigration,* partly seasonal, partly permanent and therefore have caused depopulation and abandonment of settlements. Farms were abandoned or changed into Alpine pastures. These processes are most intensive in the Alps, but they have been studied by German human geographers in Southern Europe and in the Near East as well.

On other hand the high mountains of the world have become more and more centres for *leisure and recreation,* initially for summer vacations and mountaineering, and during the last decades predominantly for winter sport activities. In this field exemplary methodological studies have been conducted by Karl RUPPERT (since 1962) and Erwin GRÖTZBACH (since 1963). German geographers have contributed substantially to the discussion of problems of regional planning in mountainous countries.

7. Comparative geography of high mountains

The great interest devoted to the geography of high mountains during the last decades is documented by the many congresses and scientific meetings which were devoted to this topic. The publications that resulted from these congresses have demonstrated that only physical geography has reached a methodically sound system, based on the pioneering work of Carl Troll. Meanwhile the human geography of high mountains is more or less at the beginning of systematic research and is searching for a common programme.

It is understandable that in this context acceptable *definitions* have to be discussed. In the word *"Hochgebirge"* the German language has a term which covers a far wider meaning than the term "hohes Gebirge" (high mountains). "Hochgebirge" is linked with certain concepts and ideas of *natural attributes.* In 1941 Carl TROLL defined as "Hochgebirge" a type of relief which exhibits glacial features, rises above the timberline and above the lower limit of periglacial solifluction into the subnival region. To this definition Carl RATHJENS (1968) und Peter HÖLLERMANN (1973) added later on the aspects of strong young uplift and rapid erosion, a pronounced vertical zonation with narrow geoecological belts caused by altitude, relief, slope angle and exposure to climatic agents. But it is doubtful whether definitions of this kind are sufficient in every case for a clear terminological differentiation.

Much greater difficulties are met by *Human Geography* in its attempts to find definitions and a typology of its topics. High mountain is a term of physical geography and may also be used as a basis for investigations by human geography without any modification. There are special types of utilization by man in all mountains of the world, and permanent settlements with agriculture have an upper limit everywhere. Therefore in general it may be sufficient to treat mountains as objects of comparisons in human geography.

Here, too, vertical zonation could be a decisive criterion. Erwin GRÖTZBACH (1980) has drawn attention to the existence and utilization of a vertical zone of Alpine pastures above the upper limit of permanent agriculture in high mountains.

In view of the increasing importance of high mountains for the recreational activities of industrial society, especially in winter sports, but also for mining, water supply and hydroelectrics, definitions based on traditional utilizations are becoming more and more dubious. But taken all in all cultural aspects of high mountains (Christoph JENTSCH 1984) are creating distinct features which are clearly differentiated from the surrounding mountain forelands by their types of settlement and economics.

Comparative geography of high mountains is based on the research work of Carl TROLL who regrettably was not able to publish his system as a comprehensive consistent programme. In his monographs about individual problems the comparison between the tropics and nontropical latitudes is stressed again and again, as is the contrast of diurnal and seasonal climates, forms of solifluction, growth patterns of vegetation and so on. Numerous symposia, mostly published in the series of "Erdwissenschaftliche Forschung", have brought together a considerable amount of information, but no attempt has yet been made to synthesize it into a comparative approach or programme (Carl RATHJENS 1980, 1982).

For instance the high mountains of Mediterranean regions, of the continental summer-dry subtropics or of monsoon climates are easily distinguished by their *physical qualities.* Altitude, altitudinal difference and characteristics of the zone of periglacial solifluction between upper timber line and snow line have a special significance in this respect. However, a first beginning of a *classification of cultural aspects* of high mountains was made at a symposium at Eichstätt (Erwin GRÖTZBACH and Gisbert RINSCHEDE 1984). There are many variables to be brought into a classification such as kinds of utilization of the zone of Alpine pastures, the vertical zonation of settlements and agriculture often ethnically differentiated, the degree of subsistence economy, of trade and of seasonal migrations of labour, the isolation and constancy of social and ethnic groups, the patterns of population densities and of emigration.

There exists already a series of separate studies investigating topics of this kind. It is probable that German high mountain research will in future pay increasing attention to such questions. Considering the change threatening nature and traditional forms of human life and economic behaviour in the high mountains of the world a rapid inventory and analysis is urgently required.

Bibliography

I. Publications with voluminous Lists of Literature

GRÖTZBACH, E. AND G. RINSCHEDE (ed.): Beiträge zur vergleichenden Kulturgeographie der Hochgebirge.
Regensburg 1984. (Eichstätter Beitr., Abt. Geogr. 12).

HARTKE, W. AND K. RUPPERT: Almgeographie.
Wiesbaden 1964. (Forschungsberichte 4 der Deutschen Forschungsgemeinschaft).

JENTSCH, CH. AND H. LIEDTKE (ed.): Höhengrenzen in Hochgebirgen.
Saarbrücken 1980. (Arb. aus dem Geogr. Inst. d. Univ. des Saarlandes 29).

LAUER, W. (ed.): Natur und Mensch in den Ökosystemen tropischer Hochgebirge.
Wiesbaden 1984. (Erdwiss. Forschg. 8).

MACHATSCHEK, F.: Das Relief der Erde.
2 vols. Berlin 1938, 1940. 2 ed. Berlin 1955.

RATHJENS, C.: Geographie des Hochgebirges.
Vol. I. Der Naturraum. Stuttgart 1982. (Teubner Studienbücher Geographie).

RATHJENS, C., C. TROLL AND H. UHLIG (ed.): Vergleichende Kulturgeographie der Hochgebirge des südlichen Asien.
Wiesbaden 1973. (Erdwiss. Forschg. 5).

TROLL, C. (ed.): Landschaftsökologie der Hochgebirge Eurasiens.
Wiesbaden 1972. (Erdwiss. Forschg. 4).

TROLL, C. AND W. LAUER (ed.): Geoökologische Beziehungen zwischen der temperierten Zone der Südhalbkugel und den Tropengebirgen.
Wiesbaden 1978. (Erdwiss. Forschg. 11).

UHLIG, H. AND W. HAFFNER (ed.): Zur Entwicklung der vergleichenden Geographie der Hochgebirge.
Darmstadt 1984. (Wege der Forschung 223).

II. Cited bibliographical data

ABELE, G.: Bergstürze in den Alpen.
München 1974. (Wiss. Alpenvereinshefte 25).

BARSCH, E.: Studien und Messungen an Blockgletschern in Macun, Unterengadin.
Z. f. Geomorph.. N. F. Suppl. vol. 8 (1969), p. 11–30.

BEUERMANN, A.: Fernweidewirtschaft in Südosteuropa.
Braunschweig 1967.

CREDNER, W.: Siam – Das Land der Tai.
Stuttgart 1935.

FINSTERWALDER, R.: Die Gletscher des Nanga-Parbat.
Z. f. Gletscherkunde 25 (1937), p. 57–108.

GRÖTZBACH, E.: Der Fremdenverkehr in den nordwestlichen Kitzbüheler Alpen.
M.G.G. München 48 (1963), p. 59–106.

GRÖTZBACH, E.: Beobachtungen an Blockströmen im afghanischen Hindukusch und in den Ostalpen.
M.G.G. München 50 (1965), p. 175–201.

GRÖTZBACH, E.: Überlegungen zu einer vergleichenden Kulturgeographie altweltlicher Hochgebirge.
Verh. 40. D.G. Innsbruck 1975. Wiesbaden 1976, p. 109–120.

GRÖTZBACH, E.: Die Nutzung der Hochweidestufe als Kriterium einer kulturgeographischen Typisierung von Hochgebirgen.
Arb. aus dem Geogr. Inst. d. Univ. des Saarlandes 29 (1980), p. 265–277.

HAFFNER, W.: Nepal – Himalaya. Untersuchungen zum vertikalen Landschaftsaufbau Zentral- und Ostnepals.
Wiesbaden 1979. (Erdwiss. Forschg. 12).

HEINE, K.: Beobachtungen und Überlegungen zur eiszeitlichen Depression von Schneegrenze und Strukturbodengrenze in den Tropen und Subtropen.
Erdk. 31 (1977), p. 161–177.

HENNING, I.: Geoökologie der Hawaii-Inseln.
Wiesbaden 1974. (Erdwiss. Forschg. 9).

HÖLLERMANN, P.: Beiträge zur Problematik der rezenten Strukturbodengrenze. Göttinger Geogr. Abh. 60 (1972), p. 135–260.

HÖLLERMANN, P.: Some reflections on the nature of high mountains, with special reference to the Western United States.
Arctic and Alpine Research 5 (1973), p. 149–160.

HÖLLERMANN, P.: Studien zur aktuellen Morphodynamik und Geoökologie der Kanareninseln Teneriffa und Fuerteventura.
Göttingen 1982. (Abh. d. Akad. d. Wiss. in Göttingen, Math.-Phys. Kl., Folge 3, No 34).

HÖLLERMANN, P.: Blockgletscher als Mesoformen der Periglazialstufe.
Bonn 1983. (Bonner Geogr. Abh. 67).

HÖVERMANN, J.: Über Strukturböden im Elburs (IRAN) und zur Frage des Verlaufs der Strukturbodengrenze.
Z. f. Geomorph., N.F. 4 (1960), p. 173–174.

HOLTMEIER, F.-K.: Die Verbreitung der Holzarten im Oberengadin unter dem Einfluss des Menschen und des Lokalklimas.
Erdk. 21 (1967), p. 249–258.

HOLTMEIER, F.-K.: Die bodennahen Winde in den Hochlagen der Indian Peaks Section (Colorado Front Range).
Münstersche Geogr. Arb. 3 (1978), p. 3–47.

HÜTTEROTH, W.-D.: Bergnomaden und Yaylabauern im mittleren kurdischen Taurus.
Marburg 1959. (Marburger Geogr. Schriften 11).

HUMBOLDT, A. v.: Voyage aux régions équinoctiales du Nouveau Continent, fait en 1799, 1800 . . .
Paris 1805–1834.

JENTSCH, CH.: Das Nomadentum in Afghanistan.
Meisenheim 1973. (Afghanische Studien 9).

JENTSCH, CH.: Methodische Ansätze in der vergleichenden Geographie der Hochgebirge (mit Beispielen aus altweltlichen Gebirgen).
Eichstätter Beitr., Abt. Geogr. 12 (1984), p. 57–72.

KLAER, W.: Untersuchungen zur klimagenetischen Geomorphologie in den Hochgebirgen Vorderasiens.
Heidelberg/München 1961. (Heidelberger Geogr. Arb. 11).

KLEINERT, CH.: Siedlung und Umwelt im zentralen Himalaya.
Wiesbaden 1983. (Geoecological Research 4).

KREBS, N.: Die Ostalpen und das heutige Österreich.
Stuttgart 1928.

KUHLE, M.: Beiträge zur Quartärmorphologie SE-Iranischer Hochgebirge. Die quartäre Vergletscherung des Kuh-i Jupar.
Göttingen 1976. (Göttinger Geogr. Abh. 67).

KUHLE, M.: Der Dhaulagiri- und Annapurna-Himalaya. Ein Beitrag zur Geomorphologie extremer Hochgebirge.
2 vols. Berlin/Stuttgart 1982. (Z. f. Geomorph., N.F. Suppl. vol. 41).

LAUER, W.: Klimatische und pflanzengeographische Grundzüge Zentralamerikas.
Erdk. 13 (1959), p., 344–354.

LAUER, W.: Zusammenhänge zwischen Klima und Vegetation am Ostabfall der mexikanischen Meseta.
Erdk. 27 (1973), p. 192–213.

LAUER, W.: Klimatische Grundzüge der Höhenstufung tropischer Gebirge.
Verh. 40. D.G. Innsbruck 1975. Wiesbaden 1976, p. 76–90.

LAUER, W.: Die hypsometrische Asymmetrie der Paramo-Höhenstufe in den nördlichen Anden.
Innsbrucker Geogr. Studien 5 (1979), p. 115–130.

LAUTENSACH, H.: Der geographische Formenwandel. Studien zur Landschaftssystematik.
Bonn 1952. (Coll. Geogr. 3).

LÖFFLER, E.: Beobachtungen zur periglazialen Höhenstufe in den Hochgebirgen von Papua New Guinea.
Erdk. 29 (1975), p. 285–292.

MONHEIM, F.: Agrargeographie der westlichen Alpen.
Gotha 1954. (P.M., Erg.-Heft 252).

PENCK, A. AND E. BRÜCKNER: Die Alpen im Eiszeitalter.
3 vols. Leipzig 1901–1909.

PILLEWIZER, W.: Untersuchungen an Blockströmen der Ötztaler Alpen.
Abh. Geogr. Inst. Freie Univ. Berlin 5 (1957), p. 37–50.

POSER, H. (ed.): Formen, Formengesellschaften und Untergrenzen in den heutigen periglazialen Höhenstufen der Hochgebirge Europas und Afrikas zwischen Arktis und Äquator.
Göttingen 1977. (Abh. d. Akad. d. Wiss. in Göttingen, Math.-Phys. Kl., Folge 3, no 31).

RATHJENS, C.: Ein Beitrag zur Frage der Solifluktionsgrenze in den Gebirgen Vorderasiens.
Z. f. Geomorph., N.F. 9 (1965), p. 35–49.

RATHJENS, C.: Neuere Entwicklung und Aufgaben einer vergleichenden Geographie der Hochgebirge.
Geogr. Taschenbuch 1966/69. Wiesbaden 1968, p. 199–210.

RATHJENS, C.: Vergleichende Geographie der Hochgebirge, an Beispielen aus den Subtropen.
Arb. aus dem Geogr. Inst. d. Univ. des Saarlandes 29 (1980), p. 15–27.

RATHJENS, C.: Vergleich und Typenbildung in der Hochgebirgsforschung.
G.Z., Beihefte, Erdkundl. Wissen 59 (1982), p. 1–8.

RINDSCHEDE, G.: Die Wanderviehwirtschaft im gebirgigen Westen der USA und ihre Auswirkungen im Naturraum.
Regensburg 1984. (Eichstätter Beitr., Abt. Geogr. 10).

RUPPERT, K. Tegernseer Tal. Sozialgeographische Studien im oberbayerischen Fremdenverkehrsgebiet.
Kallmünz/Regensburg 1962. (Münchner Geogr. Hefte 23).

RUPPERT, K.: Almwirtschaft und Fremdenverkehr in den Bayerischen Alpen.
Verh. 34. D.G. Heidelberg 1963. Wiesbaden 1965, p. 325–334.

SCHWEINFURTH, U.: Die horizontale und vertikale Verbreitung der Vegetation im Himalaya.
Bonn 1957. (Bonner Geogr. Abh. 20).

SCHWEINFURTH, U.: Neuseeland. Beobachtungen und Studien zur Pflanzengeographie und Ökologie der antipodischen Inselgruppe.
Bonn 1966. (Bonner Geogr. Abh. 36).

SCHWEIZER, G.: Büsserschnee in Vorderasien.
Erdk. 23 (1969), p. 200–205.

SCHWEIZER, G.: Nordost-Azerbaidschan und Shah Sevan-Nomaden.
G.Z., Beihefte, Erdkundl. Wissen 26 (1970), p. 81–148.

SCHWEIZER, G.: Klimatisch bedingte geomorphologische und glaziologische Züge der Hochregion vorderasiatischer Gebirge (Iran und Ostanatolien).
Erdwiss. Forschg. 4. Wiesbaden 1972, p. 221–236.

SIEGER, R.: Zur Geographie der zeitweise bewohnten Siedlungen in den Alpen.
G.Z. 13 (1907), p. 361–369.

STINGL, H. AND K. GARLEFF: Gletscherschwankungen in den subtropisch-semiariden Hochanden Argentiniens.
Z. f. Geomorph., N.F. Suppl. vol. 30 (1978), p. 115–131.

TROLL, C.: Der diluviale Inn-Chiemsee-Gletscher.
Forschg. z. dt. Landes- u. Volkskunde 23 (1924), p. 1–121.

TROLL, C.: Das Pflanzenkleid des Nanga Parbat.
Wiss. Veröffentl. d. Dt. Museums f. Länderkunde in Leipzig, N.F. 7 (1939), p. 151–180.

TROLL, C.: Studien zur vergleichenden Geographie der Hochgebirge der Erde.
Ber. Ges. d. Freunde u. Förderer der Univ. Bonn (1941), p. 49–96.

TROLL, C.: Büsserschnee (Nieve de los penitentes) in den Hochgebirgen der Erde.
Gotha 1942. (P.M., Erg.-Heft 240).

TROLL, C.: Strukturböden, Solifluktion und Frostklimate der Erde.
Geol. Rundschau 34 (1944), p. 545–694.

TROLL, C.: Die Formen der Solifluktion und die periglaziale Bodenabtragung.
Erdk. 1 (1947), p. 162–175.

TROLL, C.: Die Lokalwinde der Tropengebirge und ihr Einfluss auf Niederschlag und Vegetation.
Bonner Geogr. Abh. 9 (1952), p. 124–182.

UHLIG, H.: Typen der Bergbauern und Wanderhirten in Kaschmir und Jaunsar-Bawar.
Verh. 33. D.G. Köln 1961. Wiesbaden 1962, p. 212–222.

UHLIG, H.: Der Reisbau im Himalaya.
Erdwiss. Forschg. 5. Wiesbaden 1973, p. 77–104.

UHLIG, H.: Die Darstellung von Geo-Ökosystemen in Profilen und Diagrammen als Mittel der vergleichenden Geographie der Hochgebirge.
Eichstätter Beitr., Abt. Geogr. 12 (1984), p. 93–152.

WEISCHET, W.: Der tropisch-konvektive und aussertropisch-advektive Typ der vertikalen Niederschlagsverteilung.
Erdk. 19 (1965), p. 6–14.

WISSMANN, H. v.: Das Mitter-Ennstal.
Forschg. z. dt. Landes- u. Volkskunde 25 (1927), p. 1–144.

WISSMANN, H. v.: Süd-Yünnan als Teilraum Südostasiens.
Heidelberg 1943 (Schriften z. Geopolitik 22).

German geographical research in the Polar Regions

by

Hans Becker (Bamberg)

In 1978 the Federal Republic of Germany joined the International Antarctic Treaty of 1961. This was seen by the public as signalling a new start to German Polar Research after a long interval of inactivity. At last it seemed it was possible to continue the traditions of German Polar Expeditions established by such famous explorers as *Erich von Drygalski, Wilhelm Filcher* or *Alfred Wegener.* The foundation of the *"Alfred Wegener Institut für Polarforschung"* in 1980 at Bremerhaven was an important step in this direction. This was followed by the construction of a Polar Research vessel, the "Polarstern", put into service in 1982, and the establishment of a permanent German research station in the Antarctic, the *"Georg von Neumeyer Station"* at the eastern margin of the Weddel Sea. The break in German Antarctic research which seemed to have set in after the last expedition in 1938/39 under Alfred Ritscher to Neuschwabenland appeared to have been overcome.

There can be no doubt that the joining of the Antarctic Treaty by the Federal Republic has given a new and strong stimulus to geoscientific research in Polar regions and of course Antarctic research benefitted most by the considerable improvement in infrastructure. However, the widespread belief that all German Polar research suffered this break enforced by the war and post war events is simply not true. In fact there has been a considerable continuity of research stretching from the pre-war period to the post war years. A few examples from geographic research illustrate this:

- For several decades geographers from the University of Göttingen under the guidance of *Hans Poser* studied processes of periglacial solifluction and valley formation in the Arctic. These studies partly originated from the pre-war period and were continued in the sixties by students of Poser like

Ekkehard SCHUNKE, Kuno PRIESNITZ, Helmut STINGL and Karsten GARLEFF and by now the third generation of Poser students is working in polar and subpolar areas (LIEBRICHT 1983, 1985, RICHTER 1981). Spitzbergen, Iceland and Greenland have been the regional centres of research since Poser's pre-war expeditions. Northern Scandinavia and the Canadian Arctic were added after 1945 by GARLEFF (1970, 1984), PRIESNITZ (1983) and SCHUNKE (1981a, 1985).

- Directly before World War II *Fritz Bartz* spent several years in the USA. He stayed in Alaska for nine months where he did field work and collected material for a regional geography of Alaska. The war events delayed the publication of his results until 1950, when his extensive research went into print. Some 30 years later one of Bartz's students continued the Polar studies with a Ph.D. thesis on the Eskimos of the Keewatin region, Canadian Central Arctic (PLATZER 1980).
- The personality of *Wilhelm Dege* in particular symbolizes the continuity of German Polar research. In the thirties as a young student he undertook two research trips to Spitzbergen. The results of these studies were published in his Ph.D. thesis (W. DEGE 1938). During World War II he was able to revisit Spitzbergen in autumn 1944 when he was posted to the Rijp Fjord on the Northeastland as the leader of a military weather observation team. He collected a considerable amount of material quite apart from his meteorological observations and published these in a number of papers after the war (DEGE and BAUMANN 1960). In 1963 he conducted an extensive field trip to western Greenland mainly devoted to aspects of human geography (DEGE 1965a). Within these 30 years from 1936 till 1965 Dege published some 30 papers and books about the Arctic ranging from meteorology, climatology, geomorphology, and glaciology to human geography. Even during those difficult years between 1940 and 1950 he managed to publish 8 significant papers.
- The events of the war also influenced the research career of *Julius Büdel.* As leader of the ice forecasting service of the German Navy he participated in numerous intelligence flights across the Arctic. His "Atlas der Eisverhältnisse des Nordatlantischen Ozeans" (Atlas of the ice conditions of the North Atlantic Ocean) (BÜDEL 1950a) resulted from this work. However, much more important for Büdel's scientific development was the knowledge he gained during these flights about the Polar landscapes. This led him to distinguish within the icefree Polar region a tundra zone and a zone of frost debris (BÜDEL 1948). This subdivision of the arctic periglacial zone became incorporated into his early ideas on climatic geomorphology (BÜDEL 1950b) which he later developed into his well known concept (BÜDEL 1977, 1982). In order to verify his ideas Büdel led three expeditions to southeastern

Spitzbergen in 1959, 1960 and 1967. Some of his team members, in particular Arno Semmel and Gerhard Stäblein, carried on this work and undertook their own expeditions into the Polar regions.

The main phases and research areas of geomorphological Polar research

"Die Endmoräne des Green Bay Gletschers auf Spitzbergen – eine Studie zum Verständnis norddeutscher Diluvialgebilde" (The endmoraine of the Green Bay glacier in Spitzbergen – an investigation that permits a better understanding of the glacial features of northern Germany) is the title of a publication in 1926 by Karl GRIPP and Emmy Mercedes TODTMANN. The title and in particular the subtitle illustrate well the emphasis of German Polar research during this first period. The main rationale for the work in the Arctic was to investigate the actual periglacial processes there, in order to explain corresponding *fossil features in Central Europe.* The aim of geomorphological work in Polar regions therefore was not the exploration of these regions but the investigation of fossil and relict features outside Polar latitudes.

This emphasis did not remain uncommon after World War II. The investigation of the recent processes of gelifluction in an area where the "ice age still persists" was a major aim of Büdel's Spitzbergen expeditions. Even in his 1974 publication "Spitzbergen als Beispiel eiszeitlicher Reliefentwicklung in Mitteleuropa" (Spitzbergen as an example of Pleistocene landform development in Central Europe) this general theme still persists. However in the meantime, the emphasis of research had already shifted to the exploration and investigation of *Arctic and Antarctic landforms for their own sake.* Not the explanation of fossil landforms outside the polar regions but the study of polar-specific processes was at the centre of research. The various field investigations by the research teams of Dietrich Barsch and Gerhard Stäblein such as for instance the Heidelberg Ellesmere Island expedition (BARSCH and KING 1981), are examples for this. The research of Franz-Dieter MIOTKE (1979, 1981, 1982) in the icefree dry valleys of the Antarctic and a number of investigations by Göttinger geographers in Iceland and Greenland (STINGL and HERRMANN 1976, SCHUNKE 1977, 1979, 1980, 1983, 1985) have to be mentioned in this context.

The research into the *typical micro landforms* of the Polar regions has always been of main interest to geomorphologists, since these landforms are closely linked with the specific Polar climate, in particular the change of frost and thaw. The highly variable phenomena of patterned ground and other features of periglacial solifluction such as pingoes, palsas, thufur have been studied many times. The detailed investigations into these features and their formative processes show that they are not only the result of the climatic conditions but of a whole range of local morphodynamic factors. This has led to a new phase of German geomorphological Polar research which concentrated not so much on individual features but on the entire periglacial environment. Hans POSER (1977) called this "ecological geomorphology".

Examples of this new *ecological approach* in periglacial areas are hydrological investigations by Wolfgang-Albert FLÜGEL (1981) on Ellesmere Island, by Klaus RICHTER (1981) in Central Iceland, and by Günter NAGEL (1979) in West Spitzbergen and on Axel Heiberg Island. Dietbert THANNHEISER (1975, 1976) studied vegetation-geographical aspects in the west Canadian Arctic Archipelago and Gerhard STÄBLEIN (1977c) published papers on pedological problems in western Greenland. Investigations into *climatic factors governing soil frost conditions,* however, were the main themes of the publications, which is understandable in view of their dominating influence on periglacial morphodynamics. Gerhard STÄBLEIN (1970, 1977a, 1983a, 1985) studied the distribution and subdivision of permafrost phenomena while Ekkehard SCHUNKE and Helmut STINGL (1973) and Helga LIEBRICHT (1985) investigated the amplitude of soil temperatures on Iceland.

These authors proved that there are considerable differences between atmospheric and soil climate and that there exists no significant correlation between the two frost regimes. Other researchers devoted themselves to measurements of temperature variations inside rocks. A knowledge of these variations is important for the understanding of physical weathering in Polar regions and for the explanation of dry debris movements, the so-called dry solifluction under extremely cold arid conditions. Franz-Dieter MIOTKE (1979, 1982) conducted his research in the dry ice-free valleys of Antarctic South Victoria Land and Horst EICHLER (1981) carried out his work during the Heidelberg Ellesmere Island expedition.

As a review of the literature indicates, geomorphological research of German geographers in the Polar regions has been dominated by investigations into the processes of present day periglacial micro features. Studies of more complex *poly-genetic macro landforms* have been rather rare. Some examples are the investigations of Alfred WIRTHMANN (1964) into the pre-glacial relief generations in southeastern Spitzbergen, the geomorphological survey of Dietrich BARSCH (1981a) on the Oabloyah Bay area on Ellesmere Island and of Matthias KUHLE (1983) on West Spitzbergen.

As important as these contributions may be, the aim of this review is to concentrate on those aspects that are typical of polar regions and these are *features connected with soil frost dynamics.* Besides micro landforms such as patterned ground and other slope micro features there are also typical Polar processes of slope and valley formation. These have been emphasized by Julius BÜDEL (1969, 1972) and led him to formulate his concept of the "Eisrindeneffekt" (effect of the ice crust) and the "excessive Talbildung" (excessive valley formation) in the ice-free Polar regions.

This concept assumes that the rivers in the Arctic are characterized by a high degree of lateral erosion as well as a high capability of downward erosion termed "excessive" by Büdel. Lateral erosion is caused by the braiding streams which form due to the high sediment load of the rivers as a result of the continuous transport of debris from the highly mobile marginal slopes. The surprisingly rapid downward cutting of rivers – Büdel estimates for southeastern Spitzbergen an average rate of one to three metres per 1000 years – is explained by a mechanism which Büdel called the "Eisrinden Effekt". According to Büdel the upper zone of the permafrost near the surface contains a particularly high amount of ice (this is the ice crust = Eisrinde), and here the bedrock is fractured and broken up. This material remains frozen at first but during the following summer the relatively warm water of the rivers melts the ice crust and the rock debris is incorporated into the bedload and transported downstream. The downward cutting of the rivers in the Polar periglacial region therefore is, according to Büdel, not due to the abrasive action of the bedload but to the combination of melting and fluvial transport. Büdel claimed that this process represents the main process for valley deepening in the entire ice-free region of the Arctic.

This theory, although quite plausible, has not been substantiated by later research. Erhard BIBUS, Günter NAGEL and Arno SEMMEL (1976) or Gerhard STÄBLEIN (1977b) showed that in the valleys of Spitzbergen and Greenland no downward cutting occurs, instead accumulation clearly predominates. Dietrich BARSCH (1981b) also was not able to support Büdel's theory during his field studies on Ellesmere Island. The same applies to the work of Ekkehard SCHUNKE (1985) who compared valley forming processes in the European and American Arctic and who claims that there is no special process attributable to the permafrost that would cause excessive downward cutting. Obviously, Büdel's general model has to be replaced by much more differentiated approaches and analyses. These experiences led Dietrich BARSCH (1984) to ask for a geoecological regionalisation of the Polar areas which could form the basis of more realistic geomorphological models.

Research into the human geography of the Arctic

As an area for human settlement and economic activity the Arctic is only partly used along its margins while such activities are completely absent from the Antarctic. It is evident therefore that the main research in Polar areas is carried out within the realms of physical geography. Nevertheless, despite these preconditions, there are a number of studies in human geography with a variety of topics.

The older publications are largely devoted to a *regional description of the human geography* of the Polar areas (BLÜTHGEN 1935, 1936, 1938; DEGE 1939). This has been continued by more recent publications such as the papers of GLASER (1972) and TREUDE (1976) and by comprehensive contributions to the regional geography of Polar areas by STÄBLEIN (1977/78, 1982, 1983) or Ewald GLÄSER and Joachim SCHNÜTGEN (1986) on Iceland. Other papers fit into the category of research on time-related *structural changes* and their influence on the cultural landscape (DEGE 1964, 1965a; HEGELS 1985; PLATZER 1980; TREUDE 1974). The few studies on the historical geography of Polar areas belong to the same category (BECKER 1975; HABRICH 1972; MECKING 1949). The newest trend in research is the attempt to verify *general principles of cultural geographical processes* in the high latitudes. The author of this review has tried to demonstrate how some general aspects and mechanisms of cultural geographical processes and their dependent variables can be applied to the northern fringe of the ecumene (BECKER 1977). This concept is confirmed by TREUDE (1982) in an investigation on the development of the Eskimo economy.

A view of the investigated topics is more telling than the grouping of the individual contributions to the various research directions. As is to be expected there is no topic in German Polar geography that is specifically German. The German research into the human geography of the Arctic is closely linked with similar lines of research conducted by other countries and of course the research work of the countries fringing the Arctic have maintained a dominating role.

The review of the contribution of German geographers also shows that there is an uneven spread of topics. While some themes have been worked over frequently others have hardly been dealt with. One of these topics is *urban geography*. Apart from the two papers on the subarctic city of Workuta by Adolf KARGER (1959) and by Helmut SCHAEFER (1957) which are based on personal experience there is hardly a contribution that deals with more than some singular aspects of urban geography. Henning GRABOWSKI (1974) discusses the problems of urban growth in Barrow and Kotzebue in Alaska which have resulted from the influx of Eskimos while Manfred GABRIEL (1982) concerns himself with methods of assessing the area of influence of central places in the subpolar area of Northern America.

Another lack in German research is in the field of *historical geography* of the area north of the forested region. The few publications by Becker, Habrich and Mecking have already been mentioned in another context. From a German point of view this omission is regrettable since historical geography has a considerable reputation and tradition in Germany.

One of the topics that is particularly often studied is *lifestock breeding* in the Arctic. The economic significance of reindeer husbandry and other aspects of reindeer nomadism have been research topics for a long time. Karl SAPPER (1931) already studied reindeer nomadism. More recent investigations supplied some deeper insights (FINDEISEN 1958) or reported on new developments and innovations (MÜLLER-WILLE and PELTO 1971; GLÄSSER 1975, TREUDE 1975). The comprehensive publication of Ingrid HEMMER (1985) is the latest contribution to this topic. Very little attention by German geographers has been given to the moose which, like the reindeer, is suitable for domestication. Only Erhard TREUDE (1977) reported attempts to domesticate these animals in the Canadian Province of Quebec.

Another group of topics is centered around the significance of the Polar regions as *supplier of raw material* (DEGE 1939). As there exists a considerable tradition in mining in polar regions such as, for instance, coal mining in Spitzbergen and gold mining on the Seward Peninsula, Alaska, this has attracted geographers for some time (BLÜTHGEN 1936; DEGE 1965b; GLASER 1972). Rolf LINDEMANN (1974) also reported on oil and gas deposits in the Arctic. In recent times the emphasis of the contributions have changed; not so much the resources and their potential uses are considered but the complex problems of user conflicts (LÜKENGA 1981). In addition to this there are discussions on the political and legal conditions in the further exploration and use of Arctic regions (BUTZIN 1980; TREUDE 1983).

Finally, one theme to which German human geographers have devoted a particular interest has to be mentioned. It is the field research into the *structural change of traditional hunter and fisher societies* in the Arctic. Greenland and the Canadian Polar regions are the areas where most efforts were concentrated. As examplary one must again quote the research of Wilhelm Dege who on the west coast of Greenland investigated the causes and results of the transition from the traditional and partly self-sufficient economy to a modern market economy. DEGE (1964, 1965a) also characterized the important economic subregions of his study area in an examplary fashion.

The example set by Dege was followed by a number of similar investigations in the Canadian Polar region. Erhard TREUDE (1973, 1974) reported about the development of settlements and the economy in northern Labrador and in the eastern central Arctic of Canada. In the western central Arctic on Banks Island, Wulf HABRICH (1977) conducted a similar piece of research, and in the middle central Arctic northwest of Hudson Bay Matthias PLATZER (1980) investigated the cultural change of the Eskimos of the Keewatin region.

The Eskimo settlement of Repulse Bay is situated along the northwestern shore of Hudson Bay and it was here that Ludger Müller-Wille undertook his field studies on the technological and socio-economic change of the inhabitants. In connection with this he presented a cost analysis of modern hunting in this area (MÜLLER-WILLE 1979). There have been attempts before to research this but Müller-Wille managed to do this work with a standardized questionnaire covering every single household.

Individual research and multidisciplinary expeditions

Expeditions and research work in Polar regions during World War II like the work of Wilhelm Dege in Spitzbergen remained largely unknown to the public. This was partly because their military purpose precluded any early publicity and partly because later on in the early post-war years there was little interest in such achievements. Therefore the public gained the impression that it had been a long time since the last great German Polar expedition when in 1959/60 Julius Büdel led his first Stauferland Expeditions to Spitzbergen. In fact only just 20 years had passed since Alfred Ritscher's Antarctic expedition in 1938/39 to Neuschwabenland. As the history of expeditions in the high latitudes shows such a timespan was by no means unusual. The much higher frequency of Polar expeditions today is only a very recent feature.

In the public opinion Polar expeditions are often connected with the vision of highly adventurous trips through "night and ice". This explains why the impression of a long period of inactivity in German Polar research became entrenched in people's minds. It was rather the lack of spectacular highly publicized expeditions than the lack of actual research work that led to this misconception.

The number of German *geographical expeditions* into the Arctic and Antarctic has increased considerably since Büdel's first expeditions. Büdel's third Stauferland Expedition in 1967 was a major expedition by any measure and the "Heidelberg Ellesmere Island Expedition" led by Dietrich Barsch must also be called a large expedition. In addition there were a few smaller expeditions such as the Spitzbergen expedition by Arno Semmel and his team in 1973 and the Bellsund Expedition in 1968 by Ulrich Glaser and Gerhard Stäblein.

This last expedition, in which apart from the two leaders three other members took part, was carried out in co-operation with the German Alpine Association. This type of expedition is examplary for a number of expeditions which have become more frequent in recent times: individual researchers or small groups of scientists join larger expeditions from different disciplines or associate themselves with foreign expeditions and use the services of foreign stations. They benefit from the logistic support of these expeditions and stations but of course they also have to make compromises in the choice of their study area. Franz-Dieter Miotke was able to participate as guest researcher in the US Antarctic Research Program, and Dietrich Barsch and Gerhard Stäblein who conducted research into the periglacial Antarctic ecosystems benefitted from the logistic support supplied by the Chilean Station Teniente Rudolf Marsh (BARSCH et al. 1985).

In future it will be normal to link geographical research in the Antarctic with large *multidisciplinary projects.* For financial reasons alone no other type of expedition will be feasible. This certainly will not be a disadvantage. In the Arctic region, too, the increasing complexity of many research tasks will demand multidisciplinary teams of researchers. Here, however, there will also be room for the individual researcher and this is true in particular for the research into human geography which has been largely conducted on an individual basis all along. This will be possible because the research areas for human geographers are in the few settled areas and along the margins of the Arctic where access does not pose a major problem. In physical geography, however, the research objects are often far away from settlements or other manmade bases and more often they are in extreme environments. The high costs and technical difficulties of access as well as the numerous problems at the research site demand well organized large expeditions. In future there will still be the two forms of geographical field research in Polar regions side by side but there will be a much clearer disciplinary division between large multidisciplinary expeditions in which physical geographers participate and individual research trips by human geographers.

Bibliography

BARSCH, D.: Zur Geomorphologie des Expeditionsgebietes Oobloyah Bay, N-Ellesmere Island, N.W.T., Kanada.
In: D. Barsch and L. King (ed.), 1981, p. 109–122.

BARSCH, D.: Terrassen, Flussarbeit und das Modell der exzessiven Talbildungszone im Expeditionsgebiet Oobloyah Bay, N-Ellesmere Island, N.W.T., Kanada.
In: D. Barsch and L. King (ed.), 1981, p. 163–201.

BARSCH, D.: Periglaziale Hangformen im Oobloyah Valley, N-Ellesmere Island, N.W.T., Kanada.
In: H. Poser and E. Schunke (ed.): Mesoformen des Reliefs im heutigen Periglazialraum. Ber. über ein Symposium.
Abh. d. Akad. d. Wiss. in Göttingen, Math.-Phys. Kl., Folge 3, No. 35. Göttingen 1983, p. 171–181.

BARSCH, D.: Geomorphologische Untersuchungen zum periglazialen Milieu polarer Geosysteme. Z. f. Geomorph., N.F. Suppl. vol. 50 (1984), p. 107–116.

BARSCH, D. AND L. KING (ed.): Ergebnisse der Heidelberg Ellesmere Island Expedition. Heidelberg 1981. (Heidelberger Geogr. Arb. 69).

BARSCH, D. AND G. STÄBLEIN: Frostdynamik und Permafrost in eisfreien Gebieten der Antarktischen Halbinsel.
Polarforschung 54 (1984), p. 111–119.

BARSCH, D. AND R. MÄUSBACHER: Beiträge zur Vergletscherungsgeschichte und zur Reliefentwicklung der Südshetland Inseln.
Z. f. Geomorph., N.F. Suppl. vol. 61 (1986), p. 25–37.

BARSCH, D. et al.: Untersuchungen zum Periglazial auf der König-Georg-Insel, Südshetlandinseln/Antarktika. Deutsche physiogeographische Forschungen in der Antarktis, Bericht über die Kampagne 1983/84.
Bremerhaven 1985. (Ber. zur Polarforschg. 24).

BARTZ, F.: Alaska.
Stuttgart 1950. (Geographische Handbücher).

BECKER, H.: Siedlungswüstungen in Westgrönland. Die siedlungsgeographischen Konsequenzen wirtschaftlicher Instabilität am polaren Nordsaum der Ökumene.
Erdk. 29 (1975), p. 214–223.

BECKER, H.: Kulturgeographische Prozesse am polaren Nordsaum der Ökumene.
Polarforschung 47 (1977), p. 47–60.

BIBUS, E.: Geomorphologische Untersuchungen zur Hang- und Talentwicklung im zentralen West-Spitzbergen.
Polarforschung 45 (1975), p. 102–119.

BIBUS, E., G. NAGEL AND A. SEMMEL: Periglaziale Reliefformung im zentralen Spitzbergen.
Catena 3 (1976), p. 29–44.

BLÜTHGEN, J.: Der Verbreitungsraum der Eskimo.
Geogr. Wochenschrift 3 (1935), p. 758–763.

BLÜTHGEN, J.: Die Bodenschätze der Arktis, ihr Abbau und ihre Bedeutung.
Geogr. Anzeiger 37 (1936), p. 205–210.

BLÜTHGEN, J.: Lappland, eine landeskundliche Übersicht.
Z. f. Erdk. 6 (1938), p. 305–322.

BÜDEL, J.: Die klimamorphologischen Zonen der Polarländer.
Erdk. 2 (1948), p. 22–53.

BÜDEL, J.: Atlas der Eisverhältnisse des Nordatlantischen Ozeans mit Übersichtskarte der Eisverhältnisse des Nord- und Südpolargebietes.
Hamburg 2. ed. 1950. (Dt. Hydrologisches Inst. No. 2335).

BÜDEL, J.: Das System der klimatischen Morphologie. Beiträge zur Geomorphologie der Klimazonen und Vorzeitklimate.
Verh. 27. D.G. München 1948. Landshut 1950, p. 65–100.

BÜDEL, J.: Die Frostschutt-Zone Südost-Spitzbergens.
Bonn 1960. (Coll. Geogr. 6).

BÜDEL, J.: Wo die Eiszeit noch lebt.
Naturwissenschaft und Medizin 1 (1964), no. 2, p. 30–39; no. 3, p. 31–38.

BÜDEL, J.: Der Eisrinden-Effekt als Motor der Tiefenerosion in der exzessiven Talbildungszone.
Würzburg 1969. (Würzburger Geogr. Arb. 25).

BÜDEL, J.: Typen der Talbildung in verschiedenen klimamorphologischen Zonen.
Z. f. Geomorph., N.F. Suppl. vol. 14 (1972), p. 1–20.

BÜDEL, J.: Spitzbergen als Beispiel eiszeitlicher Reliefbildung in Mitteleuropa.
Jahrb. der Bayerischen Akad. d. Wissenschaften München 1974, p. 78–97.

BÜDEL, J.: Klima-Geomorphologie.
Berlin/Stuttgart 1977.

BÜDEL, J.: Climatic Geomorphology.
Princeton: University Press, 1982.

BUTZIN, B.: Selbstverwaltung in Grönland. Zwischen Modernisierungszwang und Wohlstandsdiktatur.
G.R. 32 (1980), p. 92–98.

DEGE, W. Geomorphologische Forschungen im nördlichen Andrée-Land (Nord-Spitzbergen).
Diss. Münster 1938.

DEGE, W.: Die heutige Besiedlung und Wirtschaft Spitzbergens.
P.M. 85 (1939), p. 166–171.

DEGE, W.: Landformende Vorgänge im eisnahen Gebiet Spitzbergens.
P.M. 87 (1941), p. 81–97, 113–122.

DEGE, W.: Grönland im Strukturwandel von Wirtschaft und Siedlung, aufgezeigt am Beispiel des Raumes um Julianehåb.
Erdk. 18 (1964), p. 169–189, 285–311.

DEGE, W.: Die Westküste Grönlands. Bevölkerung, Wirtschaft und Siedlung im Strukturwandel.
Dt. Geogr. Blätter 50 (1965), p. 5–212.

DEGE, W.: Bergbau in Grönland 1963.
Glückauf 101 (1965), p. 612–616.

DEGE, W. AND A. BAUMANN: Wissenschaftliche Beobachtungen auf dem Nordostland von Spitzbergen 1944–1945.
Offenbach a.M. 1960. (Ber. d. Dt. Wetterdienstes 10, no. 72).

DRYGALSKI, E. V. (ed.): Deutsche Südpolar-Expedition 1901–1903.
20 vols. Berlin 1905–1931.

EICHLER, H.: Gesteinstemperaturen und Insolationsverwitterung im hocharktischen Bereich, Oobloyah Bay, N-Ellesmere Island, N.W.T., Kanada.
In: D. Barsch and L. King (ed.), 1981, p. 441–464.

FINDEISEN, H.: Wandlungen in der lappischen Rentierwirtschaft.
G.R. 10 (1958), p. 129–136.

FLÜGEL, W.-A.: Hydrologische Studien zum Wasserhaushalt hocharktischer Einzugsgebiete im Bereich des Oobloyah-Tals, N-Ellesmere Island, N.W.T., Kanada.
In: D. Barsch and L. King (ed.), 1981, p. 311–382.

GABRIEL, M.: Methodische Probleme der Umlandbestimmung subpolarer Städte in Nordamerika.
In: H. Becker (ed.): Kulturgeographische Prozessforschung in Kanada.
Bamberger Geogr. Schriften 4 (1982), p. 233–243.

GARLEFF, K.: Verbreitung und Vergesellschaftung rezenter Periglazialerscheinungen in Skandinavien.
Göttinger Geogr. Abh. 51 (1970), p. 7–66.

GARLEFF, K.: Vegetationsveränderungen an periglazialen Kleinformen in Rondane/Norwegen als Klimaindikatoren.
Erdk. 38 (1984), p. 226–229.

GLÄSSER, E.: Aktuelle Wandlungen im Wirtschaftsleben der lappischen Minorität in Norwegen.
G.Z. 63 (1975), p. 139–149.

GLÄSSER, E. AND A. SCHNÜTGEN: Island.
Darmstadt 1986 (Wissenschaftliche Länderkunden 28).

GLASER, U.: Der Kohlebergbau auf Spitzbergen.
Würzburger Geogr. Arb. 37 (1972), p. 509–536.

GRABOWSKI, H.: Probleme städtischen Wachstums im arktischen Alaska.
Polarforschung 44 (1974), p. 165–176.

GRIPP, K. AND E. TODTMANN: Die Endmoränen des Green Bay Gletschers auf Spitzbergen. Eine Studie zum Verständnis norddeutscher Diluvialgebilde.
M.G.G. Hamburg 37 (1926), p. 43–75.

HABRICH, W.: Neue Erkenntnisse der Eskimoforschung
G.R. 24 (1972), p. 492–501.

HABRICH, W.: Das Leben in der Arktis. Ein Beitrag zum Mensch-Umwelt-Problem in der Kanadischen Arktis am Beispiel der Eskimos.
G.R. 29 (1977), p. 408–414.

HEGELS, F.: Klima und wirtschaftlicher Strukturwandel in Grönland.
Geowiss. in unserer Zeit 3 (1985), p. 45–51.

HEMMER, I.: Entwicklung und Struktur der Rentierwirtschaft in Finnmark und Troms (Nordnorwegen).
Bamberg 1985. (Bamberger Wirtschaftsgeogr. Arb. 1).

KARGER, A.: Workuta.
G.R. 11 (1959), p. 311–315.

KARTE, J.: Räumliche Abgrenzung und regionale Differenzierung des Periglaziärs.
Bochum 1979. (Bochumer Geogr. Arb. 35).

KARTE, J.: Entwicklung und gegenwärtiger Stand der deutschen Periglaziärforschung in den polaren und subpolaren Regionen.
Polarforschung 49 (1979), p. 97–115.

KING, L.: Palsen und Permafrost.
In: L. Müller-Wille und H. Schroeder-Lanz (ed.): Kanada und das Nordpolargebiet. Trierer Geogr. Studien, Sonderheft 2 (1979), p. 141–156.

KING, L.: Permafrost in Skandinavien. Untersuchungsergebnisse aus Lappland, Jotunheimen und Dovre/Rondane.
Heidelberg 1984. (Heidelberger Geogr. Arb. 76).

KOCH, J. P. AND A. WEGENER: Wissenschaftliche Ergebnisse der dänischen Expedition nach Dronning Louises Land und quer über das Inlandeis von Nordgrönland 1912–1913.
Kopenhagen 1930. (Meddelelser om Grønland 75).

KUHLE, M.: Zur Geomorphologie von S. Dickson Land (W-Spitzbergen) mit Schwerpunkt auf der quartären Vergletscherungsgeschichte.
Polarforschung 53 (1983), p. 31–57.

LIEBRICHT, H.: Das Frostklima Islands seit dem Beginn der Instrumentenbeobachtung.
Bamberg 1983. (Bamberger Geogr. Schriften 5).

LIEBRICHT, H.: Zum thermischen Bodenklima im Tiefland Islands.
Hveragerdi 1985. (Ber. a. d. Forschungsstelle Nedri Ás, Hveragerdi 45).

LINDEMANN, R.: Erdöl- und Erdgasvorkommen in der Arktis. Ein Überblick.
Polarforschung 44 (1974), p. 177–182.

LÖFFLER, E.: Macquarie Island, eine vom Wind geprägte Naturlandschaft in der Sub-Antarktis.
Polarforschung 53 (1983), p. 59–74.

LÜKENGA, W.: Die Alaska-Pipeline. Fallstudie eines Flächennutzungskonflikts am Nordrand der Ökumene.
Paderborn/München 1981. (Fragenkreise 23553).

MECKING, L.: Die grönländischen Normannensiedlungen im Licht ozeanischer Bedingungen und der heutigen Erwärmung der Arktis.
P.M. 93 (1949), p. 28–33.

MIOTKE, F.-D.: Zur physikalischen Verwitterung im Taylor Valley, Victoria-Land, Antarktis.
Polarforschung 49 (1979), p. 117–142.

MIOTKE, F.-D.: Formungsprozesse in eisfreien Gebieten des Victorialandes, Antarktis.
In: K. Giessner und H. G. Wagner (ed.): Geographische Probleme in Trockenräumen der Erde.
Würzburger Geogr. Arb. 53 (1981), p. 195–207.

MIOTKE, F.-D.: Hangformen und hangformende Prozesse in Süd-Victoria-Land, Antarktis.
Polarforschung 52 (1982), p. 1–41.

MIOTKE, F.-D. AND R. v. HODENBERG: Zur Salzsprengung und chemischen Verwitterung in den Darwin Mountains und den Dry Valleys, Victoria-Land, Antarktis.
Polarforschung 50 (1980), p. 45–80.

MÜLLER-WILLE, L.: Wirtschaftlicher und kultureller Wandel bei den Indianern und Eskimo in Nord-Kanada.
In: H. J. Niederehe and H. Schroeder-Lanz (ed.): Beiträge zur landeskundlich-linguistischen Kenntnis von Québec.
Trierer Geogr. Studien, Sonderheft 1 (1977), p. 113–125.

MÜLLER-WILLE, L.: Kostenanalyse der modernen Jagd bei den Inuit der kanadischen Zentralarktis.
In: L. Müller-Wille and H. Schroeder-Lanz (ed.): Kanada und das Nordpolargebiet.
Trierer Geogr. Studien, Sonderheft 2 (1979), p. 87–104.

MÜLLER-WILLE, L. AND P. J. PELTO: Technical change and its impact in Arctic regions. Lapps introduce snowmobiles into reindeer herding.
Polarforschung 41 (1971), p. 142–148.

NAGEL, G.: Untersuchungen zum Wasserkreislauf in Periglazialgebieten.
In: L. Müller-Wille and H. Schroeder-Lanz (ed.): Kanada und das Nordpolargebiet.
Trierer Geogr. Studien, Sonderheft 2 (1979), p. 157–178.

PLATZER, M.: Die Eskimo der Keewatin-Region. Ein Beitrag zum Kulturwandel in der kanadischen Zentralarktis.
Diss. Freiburg i. B. 1980.

POSER, H.: Beiträge zur Kenntnis der arktischen Bodenformen.
Geolog. R. 22 (1931), p. 200–231.

POSER, H.: Einige Untersuchungen zur Morphologie Ostgrönlands.
Kopenhagen 1932. (Meddelelser om Grønland 94,5).

POSER, H.: Talstudien aus Westspitzbergen und Ostgrönland.
Z. f. Gletscherkunde 24 (1936), p. 43–98.

POSER, H.: Bericht und Einführung. In: Formen, Formengesellschaften und Untergrenzen in den heutigen periglazialen Höhenstufen der Hochgebirge Europas und Afrikas zwischen Arktis und Äquator.
Abh. d. Akad. d. Wiss. in Göttingen, Math.-Phys. Kl., Folge 3, no. 31. Göttingen 1977, p. 7–17.

PRIESNITZ, K.: Fussflächen und Täler in der Arktis NW-Kanadas.
Verh. 43. D.G. Mannheim 1981. Wiesbaden 1983, p. 89–91.

RICHTER, K.: Zum Wasserhaushalt im Einzugsgebiet der Jökulsá á Fjöllum, Zentral-Island. Ein Beitrag zur Hydrologie des Periglazialraumes.
Göttingen 1981. (Göttinger Geogr. Abh. 78).

RICHTER, K. AND E. SCHUNKE: Runoff und Water Budget of the Blanda and Vatusdalsá Periglacial River Basins, Central Iceland.
Hveragerdi 1981. (Research Institute Nedri Ás, Hveragerdi, Bull. no. 34).

RITSCHER, A.: Wissenschaftliche und fliegerische Ergebnisse der Deutschen Antarktischen Expedition 1938/39.
2 vols. Leipzig 1942, Hamburg 1958.

SAPPER, K.: Die anthropogeographische Bedeutung des Renntieres.
G.Z. 37 (1931), p. 513–525, 599–608.

SCHAEFER, H.: Entstehung der subarktischen Grossstadt Workuta (Nordural).
Erdk. 11 (1957), p. 204–215.

SCHUNKE, E.: Die Periglazialerscheinungen Islands in Abhängigkeit von Klima und Substrat.
Göttingen 1975. (Abh. d. Akad. d. Wiss. in Göttingen, Math.-Phys. Kl., Folge 3, no. 30).

SCHUNKE, E.: Zur Genese der Thufur Islands und Ost-Grönlands.
Erdk. 31 (1977), p. 279–287.

SCHUNKE, E.: Rezente periglaziäre Morphodynamik auf Angmassalik Ø, Südost-Grönland.
Polarforschung 49 (1979), p. 1–19.

SCHUNKE, E.: Zur kryogenen Bodendynamik der arktischen Tundren Nordamerikas und Nordeuropas.
Polarforschung 51 (1981), p. 161–174.

SCHUNKE, E.: Abfluss und Abflussschwankungen im periglazialen Milieu.
Abh. der Braunschweigischen Wiss. Gesellschaft 31 (1981), p. 7–24.

SCHUNKE, E.: Aktuelle Palsabildung in der Subarktis und ihre klimatischen Bedingungen.
In: H. Poser and E. Schunke (ed.): Mesoformen des Reliefs im heutigen Periglazialraum. Bericht über ein Symposium.
Abh. d. Akad. d. Wiss. in Göttingen, Math.-Phys. Kl., Folge 3, no. 35. Göttingen 1983, p. 19–33.

SCHUNKE, E.: Vergleichende Talstudien im arktischen Periglazialraum Europas und Amerikas.
Göttingen 1985. (Nachrichten d. Akad. d. Wiss. in Göttingen, Math.-Phys. Kl., no. 1).

SCHUNKE, E. AND H. STINGL: Neue Beobachtungen zum Luft- und Bodenfrostklima Islands.
Geografiska Annaler 55 A (1973), p. 1–23.

SEMMEL, A.: Verwitterungs- und Abtragungserscheinungen in rezenten Periglazialgebieten (Lappland und Spitzbergen).
Würzburg 1969. (Würzburger Geogr. Arb. 26).

STÄBLEIN, G.: Untersuchungen der Auftauschicht über Dauerfrost in Spitzbergen.
Eiszeitalter und Gegenwart 21 (1970), p. 47–57.

STÄBLEIN, G.: Permafrost im periglazialen Westgrönland.
Erdk. 31 (1977), p. 271–279.

STÄBLEIN, G.: Rezente Morphodynamik und Vorzeitreliefinfluenz bei der Hang- und Talentwicklung in Westgrönland.
Z. f. Geomorph., N.F. Suppl. vol. 28 (1977), p. 181–199.

STÄBLEIN, G.: Arktische Böden Westgrönlands: Pedovarianz in Abhängigkeit vom geoökologischen Milieu.
Polarforschung 47 (1977), p. 11–25.

STÄBLEIN, G.: Grönland, ein Entwicklungsland in der Arktis.
Geogr. Taschenbuch 1977/78. Wiesbaden 1977, p. 27–66.

STÄBLEIN, G.: Grönland am Rand der Ökumene.
Geoökodynamik 3 (1982), p. 219–246.

STÄBLEIN, G.: Polarer Permafrost, klimatische Bedingungen und geomorphologische Auswirkungen.
Geoökodynamik 4 (1983), p. 227–248.

STÄBLEIN, G.: Formung von Hängen, Halden und Wänden. Beobachtungen im Bereich der Antarktischen Halbinsel.
In: H. Poser and E. Schunke (ed.): Mesoformen des Reliefs im heutigen Periglazialraum. Bericht über ein Symposium.
Abh. d. Akad. d. Wiss. in Göttingen, Math.-Phys. Kl., Folge 3, no. 35, Göttingen 1983, p. 160–170.

STÄBLEIN, G.: Antarktis und Arktis. Charakteristik und Bedeutung der polaren Landschaftsgürtel.
G.R. 35 (1983), p. 94–100.

STÄBLEIN, G.: Permafrost. Faktor des Naturraumpotentials in den kalten Randsäumen der Ökumene.
G.R. 37 (1985), p. 322–329.

STINGL, H. AND R. HERRMANN: Untersuchungen zum Strukturbodenproblem auf Island. Geländebeobachtungen und statistische Auswertung.
Z. f. Geomorph., N.F. 20 (1976), p. 205–226.

THANNHEISER, D.: Beobachtungen zur Küstenvegetation auf dem westlichen kanadischen Arktis-Archipel.
Polarforschung 45 (1975), p. 1–16.

THANNHEISER, D.: Ufer- und Sumpfvegetation auf dem westlichen kanadischen Arktis-Archipel und Spitzbergen.
Polarforschung 46 (1976), p. 71–82.

TREUDE, E.: Studien zur Siedlungs- und Wirtschaftsentwicklung in der östlichen kanadischen Zentralarktis.
Die Erde 104 (1973), p. 247–276.

TREUDE, E.: Nordlabrador. Entwicklung und Struktur von Siedlung und Wirtschaft in einem polaren Grenzsaum der Ökumene.
Münster 1974. (Westfälische Geogr. Studien 29).

TREUDE, E.: 40 Jahre Rentierhaltung im Mackenzie-Delta, N.W.T.
Polarforschung 45 (1975), p. 129–146.

TREUDE, E.: Pond Inlet, N.W.T. Beobachtungen zur Struktur eines eskimoischen Wirtschaftsraumes.
In: C. Schott (ed.): Beiträge zur Geographie Nordamerikas.
Marburger Geogr. Schriften 66 (1976), p. 153–173.

TREUDE, E.: Zur Moschusochsenhaltung in Nouveau-Québec.
In: H. J. Niederehe and H. Schroeder-Lanz (ed.): Beiträge zur landeskundlich-linguistischen Kenntnis von Québec.
Trierer Geogr. Studien, Sonderheft 1 (1977), p. 126–139.

TREUDE, E.: Nutzungswandel, Nutzungspotential und Raumnutzungskonflikte in der kanadischen Arktis.
Geoökodynamik 3 (1982), p. 247–269.

TREUDE, E.: Die Polargebiete, politisch-rechtliche Probleme ihrer Erschliessung und Nutzung.
G.R. 35 (1983), p. 126–130, 132.

WEGENER, K.: Geschichte der Expedition. Wissenschaftliche Ergebnisse der Deutschen Grönland-Expedition Alfred Wegener 1929 und 1930/31.
Leipzig 1933.

WIRTHMANN, A.: Die Landformen der Edge-Insel in Südostspitzbergen.
Wiesbaden 1964 (Ergebnisse der Stauferland-Expedition 1959/60, no. 2).

For the revision of the manuscript I like to thank my colleague Karsten Garleff of Bamberg University.

Index

ABDULFATTAH, K.: 98, 115, 121
ABELE, G.: 139, 144
ACHENBACH, H.: 98, 104, 115, 132
ADLAF → Arbeitsgemeinschaft Deutsche Lateinamerikaforschung
African Maps Series Project → Afrika-Kartenwerk
Afrika-Kartenwerk: 77, 79 f., 82-85, 106
Afrika-Verein: 86
AHMED DJEMAL PACHA: 94
AL-GENABI, HASHIM K.N.: 100, 115
AL-HUBAISHI, A.: 105, 115
Alfred-Wegener-Institut für Polarforschung: 149
Alpenverein, Deutsch-österreichischer: 133 f., 136, 156
AMUNDSEN, Roald: 12
ANDRAE, W.: 93
ANDRES, W.: 117
"Applied Geography and Development": 111
Arbeitsgemeinschaft Afghanistan: 107 f.
Arbeitsgemeinschaft Deutsche Lateinamerikaforschung (ADLAF): 59 f., 68
ARNOLD, A.: 115, 132
Assyriology: 102
AUF DER LANDWEHR, W.: 35
AYMANS, G.: 30, 40
BAADE, F.: 112
BADER, F.: 88
BÄHR, J.: 51, 54, 60, 64, 65, 66, 67, 68, 70, 85, 88
BALLAND, D.: 110
BANSE, E.: 94, 113
Bardai field research station: 105
BARRUCAND, M.: 97, 103
BARSCH, D.: 18, 139, 144, 151, 152, 153, 156, 157, 158, 159
BARSCH, E. et al: 157, 158
BARTELS, D.: 21
BARTELS, G.: 56
BARTH, Heinrich: 12, 78
BARTH, H. K.: 80, 81, 88, 107, 115
BARTZ, F.: 16, 150, 158
BAUMANN, A.: 150, 159
BAZIN, M.: 110
BECK, H.: 9, 13, 24
BECKER, H.: 115, 149, 154, 158, 159
BECKER, R.: 132
BEHNSTEDT, P.: 107, 115
BEHRMANN, W.: 16, 28, 39
BENECKE, D. W. et al. (ed.): 59, 70
BENEDICT, P.: 120
Berlin Geographical Society → Gesellschaft für Erdkunde zu Berlin
BERNARDES, L.: 57
BERNINGER, O.: 52, 56, 70
BETZLER, E.: 98, 107, 115
BEUERMANN, A.: 140, 144
"Bibliothek länderkundlicher Handbücher": 16
BIBUS, E.: 153, 158
BITTEL, K.: 95

BLANCKENHORN, Max: 14, 94, 113
BLÜTHGEN, J.: 154, 155, 158
BLUME, H.: 38, 48, 54, 55, 68, 70, 100, 115
BOBEK, H.: 96, 98, 113
BOESCH, H.: 30, 40, 113
BONINE, M. E.: 98, 115
BORCHERDT, C.: 65, 70
BORCHERT, G.: 49, 88
BORN, M.: 84, 88, 116
BORSDORF, A.: 55, 64, 66, 70, 75
BRECHTEL, R.: 116, 131
BREMER, H.: 36, 46, 80, 81, 88
BRONGER, D.: 34, 35, 43
BRÜCHER, W.: 64, 65, 67, 71
BRÜCKNER, Eduard: 15, 133, 146
BRÜNNOW, Rudolf Ernst: 104
BUCHHOFER, E.: 65, 67, 71, 117
BUCHHOLZ, H. J.: 27, 32, 38, 42, 48, 49
BÜDEL, J.: 16, 19, 81, 150, 151, 153, 156, 158, 159
BÜNSTORF, J.: 64, 71
BURGER, G.: 33, 35, 45
BURGEY, A.: 110
BUTZIN, B.: 155, 159
CAHEN, C.: 97
CECH, D.: 80, 85, 88
CHAI, Y. Ch.: 42
CHEVALLIER, D.: 103
CLAVAL, P.: 73
COLUMBUS, Christopher: 12
Conference reports and transactions of the German Geographers' Meetings → Deutscher Geographentag
CONRAD, Joseph: 11
COOK, James: 12, 13
CORVINUS, F.: 34, 44, 80, 85, 88
CREDNER, W.: 16, 28, 39, 55, 57, 71, 135, 144
CSIRO: 36
CZAIJKA, W.: 19, 57, 65, 71
CZAPEK, F.-M.: 116
DA GAMA, Vasco: 12
DAHLKE, J.: 36, 37, 46
DALMAN, G.: 94, 113
DAVIS, W. M.: 133
DEGE, E.: 32, 39, 41, 42
DEGE, W.: 150, 154, 155, 156, 159
DEIL, E.: 105, 116
DETTMANN, K.: 98, 99, 100, 104, 116
Deutsch-Türkisches Denkmalschutzkommando: 94
Deutsche Forschungsgemeinschaft: 20, 29, 58-62, 77, 79, 87, 95, 136
Deutsche Morgenländische Gesellschaft: 105
Deutscher Geographentag: 17, 19, 134
Deutsches Archäologisches Institut: 105
Deutsches Orient-Institut: 106
DFG → Deutsche Forschungsgemeinschaft
DÖRRIES, H.: 17
DOMASZEWSKI, Alfred von: 104
DOMIN, K.: 28, 39
DOMRÖS, M.: 18, 33, 35
DOSTAL, W.: 103, 108
DRESCH, J.: 110
DRESS, G.: 45
DRYGALSKI, E. v.: 14, 149, 159
DÜRR, H.: 31, 34, 35, 39, 41, 45
EHLERS, E.: 20, 24, 44, 96, 98, 99, 100, 101, 104, 107, 111, 116, 117
EHLERS, J.: 34
EHMANN, E.: 99, 117
EICHLER, H.: 152, 159
EIDT, R. C., et al.: 131
EL SHAGI, EL-SHAGI: 117
ELLENBERG, L.: 30, 40
ELSENHANS, H. (ed.): 73, 116
Emergency Foundation of German Science → Notgemeinschaft der deutschen Wissenschaft
ENGLISH, P. W.: 10
ERDMANN, C.: 37, 46
ERDMANN, K.: 112
"Erdwissenschaftliche Forschung": 136, 143
ERIKSEN, W.: 57, 63, 71
ESCHER, A.: 99, 117
Ethnological research: 21 f.
ETTINGHAUSEN, R.: 97, 111
EWALD, U.: 61, 71
Exploration and discovery: 11 f.
FABRICIUS, A. F.: 80, 88
FALKNER, F. R.: 78, 82, 88
FAR-HOLLENDER, U.: 20, 24
FAROUGHI, S.: 103
FAUTZ, B.: 37, 46, 47, 48
FEHLING, L.: 37, 47
FICHTER, E. M.: 34, 45
FICK, K. E.: 30, 40
FILCHNER, W.: 149
FINDEISEN, H.: 155, 159
FINSTERWALDER, R.: 17, 135, 144

FISCHER, H.: 21, 24
FISCHER, W.: 117
FISHER, C. A.: 10
FISHER, W. B.: 111, 117
FLOHN, H.: 82
FLÜCHTER, W.: 30, 40
FLÜGEL, W.-A.: 152, 159
FOCHLER-HAUKE, G.: 39, 57
FORSTER, Georg: 13
FRÄNZLE, O.: 18, 88
FRANKENBERG, P.: 82, 90, 98, 100, 117, 123
FRANZ, J. C.: 35, 44
French Overseas Research Organisation: 87
FRENZEL, B.: 18
FRENZEL, K.: 36, 37, 47
FREY, S.: 110
FREYTAG, U.: 35
FRICKE, W.: 80, 85, 88
FRIEDERICI, G.: 28, 39
FRIEDRICH, K.: 32, 43
Fundamental developments in Latin-America, Asia, and Africa → Grundlegende Entwicklungen in Lateinamerika, Asien und Afrika (project)
GABRIEL, E.: 96, 100, 117
GABRIEL, M.: 154, 159
GANS, P.: 66, 71
GARLEFF, K.: 138, 146, 150, 159, 163
GAUBE, H.: 99, 100, 103, 104, 117
GEBHARDT, H.: 118
Gegenwartsbezogener Orient, priority project: 104, 107
GEHRKE, U.: 100, 118
GEISLER, W.: 28, 38, 39
GELLERT, J.: 41
"Geographische Handbücher": 16, 100
Georg von Neumeyer-Station: 149
German Alpine Association → Alpenverein, Deutsch-österreichischer
German Association of Latin-America Research → Arbeitsgemeinschaft Deutsche Lateinamerikaforschung
German-Japanese research work-seminars: 29
German Research Foundation → Deutsche Forschungsgemeinschaft
Gesellschaft für Erdkunde zu Berlin: 14, 15, 16
Gesellschaft für Technische Zusammenarbeit, Eschborn: 86
GIERLOFF-EMDEN, H.-G.: 19, 56, 68, 71
GIESE, E.: 96, 98, 99, 118
GIESSNER, K.: 104, 118, 131, 132, 161
GLAESSER, H.-G.: 34, 45
GLÄSSNER, E.: 154, 155, 159
GLASER, G.: 57
GLASER, U.: 154, 155, 156, 159
GNIELINSKI, St. von: 37, 47, 80, 88
GOITEIN, S. D.: 97, 103, 111
GOLGER, O. J.: 32, 43
GOLTE, W.: 63, 64, 71
GORMSEN, E.: 51, 57, 58, 61, 65, 66, 67, 71, 72
GRABAR, O.: 97, 103, 111
GRABOWSKI, H.: 154, 159
GRADMANN, Robert: 17
GRENZEBACH, K.: 80, 85, 87, 88
GRIPP, K.: 151, 160
GROSSER, L.: 105, 118
GRÖTZBACH, E.: 96, 98, 104, 107, 108, 118, 139, 140, 141, 142, 143, 144
GROTZ, R.: 37, 47
Grundlegende Entwicklungen in Lateinamerika, Asien und Afrika (project): 60
GRUNEBAUM, G. von: 111
GTZ → Gesellschaft für Technische Zusammenarbeit, Eschborn
GUBLER, R.: 48
HAAS, H.-D.: 67, 72
HABRICH, W.: 154, 155, 160
HAFEMANN, D.: 18, 118, 132
HAFFNER, W.: 140, 144
HAGEDORN, H.: 96, 98, 119
HAGEDORN, H. et al.: 98, 119
HAGEN, H.: 52
HAHN, H.: 18, 96, 100, 119
HAILEY, LORD: 90
HALBACH, A.: 86
HARTKE, W.: 143
HASSINGER, Hugo: 16, 17
HAUSHERR, K.: 33, 43
HEBERER, T.: 41
HECKLAU, H.: 80, 84, 88
HEDIN, Sven: 14, 24
HEGELS, F.: 154, 160
Heidelberg Ellesmere Island expedition: 151 f., 156
HEINE, B.: 86
HEINE, K.: 37, 48, 61, 62, 72, 138, 144
HEINRICH, E.: 95, 113
HEINRITZ, G.: 99, 119

HEINZE, G. W.: 85, 88
HELBIG, K.: 28, 39
HELMSTÄDT, R.: 32, 43
HEMMER, I.: 155, 160
HENKE, Thaddäus: 12
HENKEL, R.: 84, 88
HENNING, I.: 137, 140, 144
HERRMANN, R.: 56, 151, 163
HERZFELD, Ernst: 93, 95, 104, 113, 114
HERZOG, R.: 119, 132
HETTNER, Alfred: 14, 52, 56, 72
HETZEL, W.: 84, 85, 89
High Altitude Geoecology (IGU): 134
HILLARY, Sir Edmund: 12
HILLENBRAND, H.: 30, 40
HIRAKAWA, K.: 40
HODENBERG, R. v.: 161
HÖHFELD, V.: 100, 115, 119
HÖLLERMANN, P.: 137, 138, 139, 140, 142, 144, 145
HÖLPER, A.: 38, 49
HÖVERMANN, J.: 19, 138, 145
HOFMEISTER, B.: 37, 47
HOHNHOLZ, J.: 33, 44
HOLTMEIER, F.-K.: 139, 145
HOPFINGER, H.: 119
HORSTMANN, K.: 34, 35, 45
HOSRY, M.: 99, 119
HOTTES, K.-H.: 127
HOURCADE, B.: 110
HUECK, K.: 57
HÜMMER, Ph.: 98, 120
HÜTTEROTH, W.-D.: 96, 98, 99, 100, 101, 107, 108, 112, 120, 121, 140, 145
HUMBOLDT, Alexander von: 8, 9, 11, 13, 14, 15, 24, 51, 95, 139, 145
IBRAHIM, F. N.: 81, 82, 89, 96, 104, 121
INALCIK, H.: 103
Institut für Afrika-Kunde: 86
Instituto Colombo-Aleman in Santa Marta: 56
Instituto Tierras y Colonización: 65
Instituto Tropical de Investigaciones Científicas (ITIC): 55 f.
International Geographical Congress: 111
International Geographical Union: 9, 54, 110, 134
Islamic studies: 97, 102, 111
ISSAWI, Ch.: 97
JAEGER, F.: 16, 17, 78, 82, 89
JÄCKEL, D. et al.: 105, 121
JEANNERET, F.: 38, 48
JÄTZOLD, R.: 80, 82, 84, 89
JANZEN, J.: 98, 99, 121, 128
JASCHKE, D.: 37, 47
JEBENS, A.: 99, 121
Jemen-Project: 107
JENTSCH, Ch.: 96, 98, 99, 108, 121, 122, 140, 142, 143, 145
JESKE, J.: 85, 89
JESSEN, O.: 78, 81, 89
JOHANSEN, U.: 103, 108
JOHNSTON, R. J.: 73
JORDAN, J.: 95
JÜLICH, V.: 64, 72
JUNGFER, E.: 122
KAEMPFER, Engelbert: 13
KAISER, F. G.: 65
KARGER, A.: 154, 160
KARPAT, K.: 97, 111
KARTE, J.: 160
KAYSER, K.: 19, 78, 79, 80, 81, 89, 91
KERSBERG, H.: 89
KESSLER, A.: 57
KIEPERT, Heinrich: 94
KIEPERT, Richard: 94
KING, L.: 151, 157, 158, 159, 160
KINZL, H.: 19
KIRCHHOFF, P.: 60
KLAER, W.: 38, 48, 98, 122, 139, 145
KLAHSEN, E.: 65, 72
KLAUS, D.: 82, 89
KLEINERT, Ch.: 141, 145
KLIMM, E.: 80, 87, 89
KLÜCKMANN, G.: 67, 70
KLUTE, Fritz: 16, 17, 25, 53, 78, 89, 114
KNAPP, R.: 83, 89
KNOCHE, W.: 52
KNOTHE, H.: 17
KOCH, J. P.: 160
KOCH, W.: 30, 41
KOCHWASSER, F. H. (ed.): 100, 122
KÖHLER, O.: 86
KÖSTER, G.: 65, 72
KOHLHEPP, G.: 56, 57, 60, 64, 67, 72, 74
KOLB, A.: 16, 19, 28, 31, 34, 39, 43, 44
KOLDEWEY, Robert: 93
KOPP, H.: 96, 98, 99, 100, 104, 107, 115, 122, 123

KORBY, W.: 123
KORTUM, G.: 123
KOSCHATZKY, K.: 34, 44
KOSWIG, C.: 112
KRÄMER, A.: 28, 39
KRAUS, W.: 98, 99, 100, 108, 119, 123
KRAUSE, R. F.: 123
KREBS, Norbert: 16, 45, 135, 145
KREISEL, W.: 38, 48, 49
KREISER, K.: 103
KRENCKER, Daniel: 104
KRETH, R.: 67
KROSS, E.: 65, 72
KUDER, M.: 80, 89
KÜCHLER, J.: 31, 32, 33, 34, 42, 43, 44, 45
KÜHN, F.: 16, 17, 52
KÜHNE, D.: 34, 43, 44, 45
KÜNDIG-STEINER, W. (ed.): 100, 123
KUHLE, M.: 31, 42, 138, 145, 152, 160
KULINAT, K.: 65, 67, 70, 72
KULKE, E.: 34, 45
KULS, W.: 83, 84, 89, 90
KURZROCK, R. (ed.): 121
LAMPING, H.: 37, 47
LAUER, W.: 18, 48, 52, 53, 54, 55, 56, 59, 60, 61, 62, 63, 72, 73, 82, 88, 90, 98, 119, 123, 137, 139, 143, 144, 145
LAUTENSACH, H.: 16, 17, 28, 39, 137, 145
LEEMANN, A.: 45, 46
LEHMANN, H.: 19, 28, 45
LEIDLMAIR, A.: 74, 96, 99, 100, 123
LEIPPERT, H.: 123, 132
LENZ, K.: 51, 58, 67, 72
LESER, H.: 18, 82, 90
LESER, H. (ed.): 90
LEUNG, Ch.-K.: 42
LEUPOLD, W.: 44
LEWIS, B.: 97, 111
LIEBRICHT, H.: 150, 152, 160
LIEDTKE, H.: 38, 48, 143
LIENAU, C.: 80, 90
LINDAUER, G.: 35
LINDEMANN, R.: 155, 160
LISS, C. C.: 63, 73
LIVINGSTONE, David: 12
LÖFFLER, E.: 36, 37, 38, 44, 47, 49, 138, 145, 160
LOOSE, R.: 122
LORCH, W. T: 35, 44
LOUIS, H.: 16, 19, 45, 81, 90, 98, 112, 114
LUDWIG, H. D.: 84, 90
LÜKENGA, W.: 155, 160
LUTZ, W.: 38, 48
MAASS, A.: 64, 73
MACHATSCHEK, Fritz: 16, 57, 134, 143
MÄUSBACHER, R.: 158
MAHNKE, L.: 63, 73
MAJER, H. G.: 131
MALIK, M. U.: 100, 123
MANSHARD, W.: 79, 80, 83, 84, 90
MANSUR, F.: 120
MARSH, R.: 157
MATZAT, W.: 33, 44
MATZNETTER, J.: 80, 84, 90
MAUCH, C.: 78
MAULL, Otto: 17, 52
MAYER, J.: 100, 123
MAYR, A.: 30, 40
MECKELEIN, W.: 96, 123
MECKING, L.: 17, 27, 39, 154, 155, 160
MEDER, O.: 123
MEHNER, H.: 100, 118
MEINECKE, M.: 103
MENSCHING, H.: 19, 79, 80, 81, 82, 90, 95, 100, 101, 104, 106, 123, 124
MERTINS, G.: 56, 60, 65, 66, 67, 70, 73, 117
MESSERLI, G.: 98, 124
MÉTRAL, F.: 131
METZ, F.: 17
METZNER, J. K.: 34, 35, 45
Mexico-Project: 60-62, 136
MEYER, G.: 98, 99, 107, 124, 125
MEYER-ABICH, A.: 55
MEYNEN, E.: 130
Middle East Studies Association (MESA): 111
MIKESELL, M. W. (ed.): 10, 24
MIKUS, W. (ed.): 67, 73
MIOTKE, F.-D.: 151, 152, 157, 160, 161
MITTENDORFF, R.: 66, 73
MOISEL, B. M.: 78
MOMENI, M.: 100, 124
MONHEIM, F.: 18, 54, 57, 64, 65, 73, 140, 145
MORALES, M.: 60, 73
MORTENSEN, H.: 17, 19
MÜLLER, J.: 68, 73
MÜLLER-HOHENSTEIN, K.: 105, 115, 116, 124, 125
MÜLLER-WIENER, W.: 103, 125

MÜLLER-WILLE, L.: 154, 156, 160, 161
NACHTIGAL, Gustav: 78
NAGEL, G: 152, 153, 158, 161
Nanga Parbat expedition (German): 134
NANSEN, Fridtjof: 12
Near Eastern archeology: 111
NICKEL, H. J.: 61, 73
NIEBUHR, Carsten: 13
NIEDEREHE, H. J.: 161, 163
NITZ, H.-J. (ed.): 120
NÖLDECKE, Arnold: 93
Notgemeinschaft der deutschen Wissenschaft: 17, 58 → Deutsche Forschungsgemeinschaft
NUHN, H.: 60, 65, 66, 67, 73, 74
NUHN, H. (ed.): 74
OBERBECK, G.: 49
OBST, E.: 17, 78, 81, 90
OBST, E. (ed.): 91
OMINDE, S. H.: 80, 85, 88
OPPENHEIM, Max Frh. von: 93, 114
O'REILLY-STERNBERG, H.: 57
ORSTOM → French Overseas Research Organisation
OTTE, F.: 27, 39
Ottoman Empire: 93 f., 99
Ottoman studies: 102
PACHNER, H.: 67, 74
PAFFEN, K. H.: 18, 19, 54
PANZER, W: 27, 45
PAPE, H.: 100, 125
PARET, R. (ed.): 114
PARSONS, J. J.: 10
PARTSCH, Joseph: 17
PASSARGE, S.: 16, 17, 39
PELTO, P. J.: 154, 161
PENK, Albrecht: 15, 16, 17, 24, 133, 134, 146
PFEIFER, G.: 56, 57, 60, 65, 74
PHILIPP, H.-J.: 99, 125
PHILIPPSON, Alfred: 14, 94, 114
PILLEWIZER, W.: 19, 139, 146
PLANCK, U.: 99, 103, 112, 125
PLANHOL, X. de: 97, 110
PLATZER, M.: 150, 154, 155, 161
PLEWE, E.: 130
Polarstern expedition: 149
POLO, Marco: 12
POPP, H.: 98, 99, 125, 126
POPP, K.: 66, 74
POSER, H.: 138, 146, 149, 152, 157, 161, 162
PRIDHAM, B. R.: 123, 124, 128
PRIESNITZ, K.: 150, 161
QUELLE, O.: 52
RAPPENHÖNER, D.: 105, 126
RATHJENS, C.: 19, 95, 96, 98, 101, 108, 118, 120, 126, 133, 138, 140, 142, 143, 146
RATHJENS sen., C.: 95, 96, 104, 114
RAYMOND, A.: 97, 103
REIN, J. J.: 27, 39
REINER, E.: 36, 37, 47
REINHARD, U.: 108
REUTER, Ernst: 112
RICHTER, G.: 80, 88
RICHTER, K.: 150, 152, 161
RICHTER, M.: 126
RICHTHOFEN, Ferdinand Frh. von: 8, 9, 11, 12, 13, 14, 15, 16, 17, 24, 27, 39, 42, 95
RIETHMÜLLER, R.: 44
RINSCHEDE, G.: 141, 143, 146
RIST, B.: 100, 126
RITSCHER, A.: 156, 161
RITTER, W.: 100, 126
RODNEY, W.: 85, 91
RÖLL, W.: 34, 35, 43, 45, 46
RÖLLIG, W.: 103, 106, 126
RÖNICK, V: 65, 74
ROHDENBURG, H.: 80, 88
ROHMEDER, W.: 55, 67, 75
ROHMER, W.: 116, 131
ROTHER, K.: 37, 47, 65, 66, 74
ROTHER, L.: 100, 127
RUPPERT, H.: 100, 104, 127
RUPPERT, K.: 140, 141, 143, 146
RUPPIN, A.: 94, 114
RUTZ, W.: 34, 44, 46
SACK, D.: 103
SANDER, H. J.: 61, 74
SANDNER, G.: 49, 60, 65, 66, 67, 73, 74
SAPPER, Karl: 16, 17, 28, 39, 52, 56, 155, 162
SARRE, Friedrich: 104, 114
SAUER, C. O.: 54, 57
SCHAEFER, J.: 154, 162
SCHAER-GUHL, S.: 70
SCHÄTZL, L.: 84, 91
SCHAMP, E. W.: 84, 91
SCHAMP, H.: 99, 100, 127
SCHARABI, M.: 98, 103, 127
SCHEIDL, L.: 28, 39
SCHERM, G.: 67, 74

SCHEU, E.: 17, 90
SCHIMMEL, A.: 100, 123
SCHLIEPHAKE, K.: 100, 127, 130
SCHLÜTER, Otto: 17
SCHMID, W.: 37
SCHMIDT-WULFFEN, W. D.: 82, 91
SCHMIEDECKEN, W.: 126
SCHMIEDER, O.: 17, 52, 53, 54, 55, 56, 63, 64, 67, 74, 75
SCHMITHÜSEN, J.: 57
SCHMITTHENNER, Heinrich: 17, 27, 39
SCHNEIDER, Ch.: 126
SCHNEIDER, H.: 65, 70
SCHNEIDER, J.: 43
SCHNEIDER, K.-G.: 80, 84, 89, 91
SCHNÜTGEN, J.: 154, 159
SCHÖLLER, P.: 18, 29, 32, 38, 39, 40, 41, 42, 43, 49
SCHOLZ, F.: 96, 98, 99, 100, 101, 104, 107, 121, 127, 128
SCHOLZ, U.: 33, 34, 35, 43, 46
SCHOOP, W.: 63, 64, 66, 75
SCHOTT, C.: 163
SCHREPFER, Hans: 17
SCHROEDER-LANZ, H.: 160, 161, 163
SCHULTZ, J.: 80, 84, 91
SCHULTZE, J. H.: 71, 78, 79, 91
SCHULZ, A.: 17, 42
SCHULZE, W.: 80, 91
SCHUMACHER, G.: 94, 114
SCHUNKE, E.: 150, 151, 152, 153, 157, 161, 162
SCHURIG, W.: 37, 48
SCHWEINFURTH, Georg: 12, 78
SCHWEINFURTH, U.: 18, 35, 36, 37, 46, 47, 48, 138, 139, 146
SCHWEIZER, G.: 96, 98, 99, 101, 107, 123, 125, 128, 137, 139, 140, 146
SCHWIND, M.: 28, 29, 41
SEDLACEK, P.: 37, 47
SEELE, E.: 61, 75
SEGER, M.: 100, 128
SELLENBERG, E.: 37, 48
SEMMEL, A.: 83, 91, 151, 153, 156, 158, 162
ŞEN, E.: 99, 128
SENFTLEBEN, W.: 33, 45
SICK, W. D.: 65, 75, 80, 91
SIEBOLD, Philipp Franz von: 12
SIEGER, Robert: 135, 146
SIEVERS, W.: 52, 67, 75
SIMKO, H. D.: 32, 43
SOURDEL-THOMINE, J.: 97
SOYSAL, M.: 98, 120, 129
SPÄTH, H.: 36, 47
SPIEKER, U.: 100, 129
SPÖNEMANN, J.: 80, 91
SPRIGADE, P.: 78
STADELBAUER, J.: 31, 42, 98, 99, 129
STÄBLEIN, G.: 151, 152, 153, 154, 156, 157, 158, 162, 163
STAGL, J.: 21, 22
Stauferland expedition (1959-60, 1967): 156
STEGER, H.-A.: 67, 74
STEIN, N.: 33, 35, 46
STEWIG, R.: 96, 100, 104, 129, 130
Stiftung Volkswagenwerk: 20, 58, 60, 104, 107
STINGL, H.: 138, 146, 150, 151, 152, 162, 163
STODDART, D. R.: 10, 11, 24
STÖBER, G.: 98, 129
STRÄTZ, C.: 105, 129
STRAUB, R.: 107, 129
STRUCK, E.: 129
STUCKMANN, G.: 104, 129, 131, 132
SUM KONG-SUT: 43
TAFEL, Albert: 14
TAUBMANN, W.: 31, 42
TAVO → Tübinger Atlas des Vorderen Orients
TERMER, F.: 17, 52, 60
THANNHEISER, D.: 152, 163
THILENIUS, G.: 28
THOMASIUS, H.: 44
THOMPSON, W. P.: 44
THORBECKE, Franz: 16
THÜNEN, Johann Heinrich von: 56
Thyssen Foundation: 136
TICHY, F.: 43, 57, 61, 75, 126
TICHY, F. (ed.): 75
TIESSEN, E. (ed.): 24
TIMMERMANN, O. F.: 30, 41
TODTMANN, E.: 151, 160
TOEPFER, H.: 98, 130
TRAUTMANN, W.: 61, 75
TREUDE, E.: 154, 155, 163
TROIN, J. F.: 97, 110
TROLL, Carl: 8, 9, 17, 18, 19, 25, 48, 52, 53, 54, 55, 62, 73, 75, 83, 118, 120, 126, 134, 135, 136, 137, 138, 139, 142, 143, 144, 146, 147

Tübinger Atlas des Vorderen Orients: 106
TÜMERTEKIN, E.: 120
TYRAKOWSKI, K.: 61, 75
UHLIG, H.: 32, 37, 43, 44, 47, 56, 118, 120, 140, 141, 143, 144, 147
UMMENHOFER, S. M.: 67, 75
UNESCO: 35
VALENTIN, H.: 36, 47
VALVREDE, O.: 57
Verhandlungen des Deutschen Geographentages → Deutscher Geographentag
Völkerkundlich-kulturgeographischer Arbeitskreis gegenwartsbezogener Orient: 108 f.
Volkswagen Foundation → Stiftung Volkswagenwerk
VORLAUFER, K.: 84, 91
VOSS, F.: 34, 35, 44
WAGNER, E.: 47
WAGNER, Hermann: 17
WAGNER, H.-G.: 98, 118, 130, 132, 161
WAHNSCHAFFT, R.: 44
WAIBEL, Leo: 16, 17, 52, 56, 57, 63, 74, 75, 78, 91
WALTER, H.: 83, 91
WATZINGER, C.: 94, 104, 114, 115
WEGENER, Alfred: 17, 149, 160, 163
WEGENER, K.: 163
WEHLING, H.-W.: 37, 47
WEIGT, E.: 79, 91
WEIN, N.: 31, 42
WEISCHET, W.: 18, 54, 56, 68, 75, 83, 91, 92, 137, 147
WENZEL, H.: 114
WERNEBURG, K.-H.: 30, 41
WICHE, E.: 19
WIDMER, U.: 31, 41, 42
WIEBE, D.: 100, 104, 130
WIEGAND, Theodor: 94
WIEMER, H.-J.: 38, 49
WIESE, B.: 77, 80, 84, 88, 89, 90, 91, 92, 106
WILHELMY, H.: 33, 38, 43, 49, 54, 55, 63, 64, 65, 66, 68, 75
WIRTH, E.: 7, 19, 21, 25, 77, 79, 93, 96, 98-101, 103 f., 107 f., 111, 113-128, 130 f.
WIRTHMANN, A.: 38, 49, 152, 163
„Wissenschaftliche Länderkunden“: 80, 100
WISSMANN, H. von: 27, 74, 78, 95, 98, 104, 114, 126, 135, 147
WOLDSTEDT, P.: 37, 38, 48
WULZINGER, K.: 94, 104, 114, 115
YAZAWA, T.: 29
ZEIDLER, H.: 123, 132
„Zeitschrift des Deutschen Palästinavereins“: 93 f.
ZIA TAVANA, M. H.: 131
ZIMMERMANN, G. R.: 34, 35, 46
ZIMPEL, H. G.: 131
ZSCHIETZSCHMANN, Willy: 104